WHAT THE FDA AND THE GOVERNMENT ARE COVERING UP

THE TOXIC INGREDIENTS THEY ARE PUTTING IN OUR FOOD!

KATHY A. YNCERA

CONTENTS

Introduction	7
1. UNVEILING THE HIDDEN	9
1.1 The Role of the FDA in Food Safety Oversight	10
1.2 Decoding the Term "Bioengineered": Implications for Consumer Health	12
2. THE SCIENCE OF SECRECY	19
2.1 Understanding Food Label Tricks and Loopholes	19
2.2 The Impact of Preservatives on Long-Term Health	22
2.3 How Food Colorings Alter Our Biology	24
3. MAJOR CULPRITS	27
3.1 Trans Fats and Heart Disease: A Hidden Epidemic	27
3.2 High Fructose Corn Syrup and Obesity	30
3.3 MSG and Its Neurological Effects	32
4. REGULATORY FAILURES	35
4.1 The Influence of Big Food Companies on FDA Policies	36
4.2 Case Study: The Aspartame Approval Controversy	39
4.3 How Lobbying Shapes Food Safety Standards	41
5. GLOBAL COMPARISONS	45
5.1 European vs. American Food Safety Standards	45
5.2 Lessons from Japan: A Case for Stricter Regulations	48
5.3 The Canadian Approach to GMOs	50
6. CONSUMER IMPACT	55
6.1 Linking Diet to Disease: What the Data Shows	55
6.2 Allergies and Additives: A Growing Concern	58
6.3 The Rise of Food Sensitivities and Intolerances	60

7. INDUSTRY INSIDERS SPEAK 63
   7.1 Confessions of a Food Scientist 63

8. IMPACT OF WHISTLEBLOWING 67
   8.1 Investigative Journalism and Food Safety 68
   8.2 Mainstream Media vs. Independent Reporting on Health Issues 71
   8.3 Social Media's Influence on Food Trends and Misinformation 74

9. LANDMARK LAWSUITS IN FOOD SAFETY 77
   9.1 Recent and Ongoing Legal Battles 79
   9.2 The Power of the Freedom of Information Act (FOIA) 80
   9.3 Advocacy Groups and Their Role in Legal Reforms 82

10. ETHICAL CONSIDERATIONS 87
    10.1 Ethical Responsibility of Food Manufacturers 87
    10.2 Consumer Rights vs. Corporate Security 90
    10.3 The Ethics of Bioengineering 92

11. ACTIVISM AND ADVOCACY 97
    11.1 Starting a Grassroots Movement for Food Safety . . 98
    11.2 Effective Lobbying Techniques for the Everyday Citizen 100
    11.3 Using Social Media for Advocacy and Change 102

12. THE FUTURE OF FOOD 105
    12.1 The Role of AI in Food Safety Monitoring 105
    12.2 The Potential of Blockchain in Tracking Food Ingredients 108
    12.3 Innovations in Non-Toxic Food Preservation 110

13. BUILDING RESILIENCE 115
    13.1 Staying Informed Without Becoming Overwhelmed 115
    13.2 Supporting Local and Organic Food Sources 117
    13.3 Mindfulness and Dietary Choices 119

14. GLOBAL MOVEMENTS   123
    14.1 The Slow Food Movement: A Global
    Perspective   124
    14.2 International Coalitions for Food Safety   126
    14.3 Case Study: The Indian Organic Food
    Revolution   129

15. CALL TO ACTION   133
    15.1 How to Read and Understand Food Labels
    Accurately   134
    15.2 Tools for Tracking and Reporting Food Safety
    Violations   136
    15.3 Organizing Community Forums on Food Safety   138
    15.4 Engaging Youth in Food Safety Advocacy   141
    15.5 Writing to Legislators: A Template for Change   143
    15.6 Future Technologies and Their Role in
    Consumer Empowerment   146

    Conclusion   149
    References   151

INTRODUCTION

Every year, the average consumer unknowingly ingests numerous substances that were never meant for human bodies. Recent **investigations have unveiled that some of these substances have slipped through regulatory nets and are lurking in our everyday** foods. This revelation is not just troubling; it is a clear and present danger to our health.

I am not a scientist or a politician. My journey into the murky waters of food safety began when a close family member fell ill from what was initially a mystery ailment. It turned out to be linked to harmful additives in seemingly safe foods. This personal crisis catapulted me into action, and I became a fervent advocate for food safety and transparency. I am driven by a simple yet profound belief that everyone has the right to know precisely what they put into their bodies.

This book is my effort to peel back the curtain on the toxic ingredients approved by the FDA and hidden in plain sight in our food. We will explore how these substances were approved, their impacts on human health, and how we can avoid them. The scope

Introduction

of this problem is vast, touching every plate and pantry with implications that stretch into future generations.

The issue of toxic ingredients in our food affects everyone—children, adults, seniors, healthy individuals, and those with chronic illnesses. It is not a niche problem; it is a universal crisis. By understanding the history of food regulation, exploring scientific findings on toxic ingredients, and reading case studies on specific harmful substances, we can start to grasp the issue's magnitude.

In the following chapters, I will share the cold, hard facts, personal stories, and case studies that bring the statistics to life. These narratives underscore the human impact of food safety failures and the urgent need for change. For instance, consider the widespread use of certain bioengineered ingredients linked to increased health risks—a fact I stumbled upon during my research, which was shocking and eye-opening.

What can you do when the very agencies meant to protect us seem to falter? As ordinary consumers, how can we fight against giant food corporations prioritizing profits over health? These are some of the critical questions we will tackle. This book is not merely an exposé; it is a call to arms, urging you to join a growing movement demanding stricter food safety regulations and greater transparency.

Together, we can advocate for a food system prioritizing public health over corporate profits. We can make informed choices about our eating and advocate for a transparent and accountable system. Join me in this fight for our health, rights, and future. Let us unite against these hidden dangers and ensure a healthier tomorrow for ourselves and future generations.

CHAPTER 1
UNVEILING THE HIDDEN

In 2015, a significant food corporation faced a severe crisis that led to a national recall of several popular processed foods, traced back to contaminated spices that had slipped through routine safety inspections. This incident is not isolated but rather a glaring example of systemic issues within the regulatory frameworks intended to safeguard our food. As you delve into the complexities of the Food and Drug Administration's (FDA) role in overseeing food safety, you will begin to understand how deeply these systemic issues run. This chapter aims to shed light on the foundational structures, the effectiveness of oversight, the undeniable impact of lobbying, and the recent changes within the FDA that affect what ends up on your dinner table.

1.1 The Role of the FDA in Food Safety Oversight

Regulatory Framework Overview

The FDA, established to protect and promote public health, regulates approximately $1 trillion worth of products annually, which accounts for about a quarter of all consumer expenditures in the United States. This immense responsibility includes ensuring the safety and efficacy of foods, pharmaceuticals, medical devices, and cosmetics. Specifically, within the food sector, the FDA's role is multifaceted, encompassing the evaluation and approval of food additives, oversight of food labeling, implementation of safety protocols for food processing, and enforcement actions against entities that violate these standards.

The FDA's regulatory framework is designed to prevent contamination and ensure that foods are safe, nutritious, and labeled accurately. This involves a series of proactive risk assessments, compliance checks, and reactive enforcement measures. The FDA also sets scientific standards for testing foods and food-related products to ensure they meet safety guidelines before being marketed to consumers. However, the breadth of the FDA's mandate and limited resources often lead to challenges in fully executing its duties.

Critique of Oversight Mechanisms

Despite the comprehensive framework, there have been notable failures in the FDA's oversight that highlight areas for improvement in its execution. For instance, the national recall mentioned earlier resulted from systematic failures to detect contamination in food additives. Such incidents raise questions about the efficacy of the FDA's inspection and compliance processes. Studies suggest that the FDA inspects less than two percent of all food imports, which may contribute to such oversights. Furthermore, the rapid increase in global food systems has stretched the FDA's capacity thin, making it difficult to adequately monitor every facet of food safety.

Moreover, the process of approving food additives has been criticized for its reliance on data from studies funded by the food industry, which can introduce biases and conflicts of interest. This reliance is concerning, given the potential for these studies to downplay or omit findings that would negatively affect product approval. These conflicts of interest, whether real or perceived, undermine public trust in the FDA's ability to protect consumers from harmful food ingredients.

Impact of Lobbying

Lobbying by the food industry has significantly influenced FDA policies, often resulting in less stringent safety checks and the approval of questionable additives. The food and beverage industries collectively spend millions on lobbying annually, aiming to shape legislation and regulation in ways that favor their business interests. For example, intense lobbying efforts have delayed implementing updated nutritional labeling designed to provide more precise sugar and calorie content information. These delays hinder consumer rights to informed choices and reflect how deeply industry interests are embedded in regulatory processes.

Recent Reforms and Changes

The FDA has initiated several reforms to strengthen food safety oversight in response to growing public and political pressure. The Food Safety Modernization Act (FSMA), enacted in 2011, represents the most significant expansion of the FDA's food safety authority in over 70 years. The FSMA has shifted the focus from responding to contamination to preventing it, requiring frequent facility inspections based on risk and mandating comprehensive, prevention-based controls across the food supply.

This shift includes new rules for the sanitary transportation of human and animal food, measures to prevent intentional adulteration, and enhanced protocols for tracing foodborne illness outbreaks back to their source more quickly. While these reforms are steps in the right direction, their implementation has been gradual, and the effectiveness of these changes in reducing food safety incidents remains to be fully seen.

As we continue to scrutinize the FDA's role, it becomes clear that while structures are in place to ensure food safety, their effectiveness is often compromised by external pressures and internal limitations. Understanding these dynamics is crucial for advocating for stronger regulations and more transparent practices prioritizing public health over industry profits.

1.2 Decoding the Term "Bioengineered": Implications for Consumer Health

In recent years, the term "bioengineered" has become a staple on food labels, nestled among other ingredients in fine print. The United States Department of Agriculture (USDA) defines bioengineered foods as those containing detectable genetic material modified through specific laboratory techniques and cannot be created through conventional breeding or found in nature. Labeling is part of an effort to increase consumer transparency about what is in their food, particularly in the context of genetically modified organisms (GMOs). However, using "bioengineered" rather than the more commonly known "GMO" in labeling might confuse consumers unfamiliar with the terminology. This subtle shift in language affects consumer perception and might lead some to believe they are avoiding GMOs when, in fact, they are consuming products that have been genetically altered.

The health implications of consuming bioengineered foods are a topic of hot debate and scientific study. Some studies suggest that these foods are safe and have no difference in nutritional value compared to their non-modified counterparts. However, other research hints at potential health risks such as increased allergenicity or the introduction of new toxins through genetic modifications. For example, a study might indicate that a new strain of bioengineered corn produces a protein that, while intended to fend off pests, could cause mild allergic reactions in a small population segment. These potential health impacts are problematic, given the gaps in regulatory policies that allow bioengineered foods to enter the market with relatively lax oversight compared to medical products. The FDA's policy on genetically engineered foods, which operates under the assumption that they do not differ fundamentally from other foods, has been criticized for not requiring mandatory safety testing before such foods are sold.

Moreover, the regulatory framework for bioengineered foods does not comprehensively address the long-term health impacts of consuming these products. Current policies focus on a pre-market review that examines the potential for allergic reactions and toxic effects from new proteins in bioengineered foods. However, this review process is voluntary for food producers. It does not require long-term health studies, leaving a significant gap in our understanding of how these foods affect health over time. This gap is particularly concerning as bioengineering becomes more sophisticated, introducing more complex modifications across a more comprehensive array of food products.

Consumer perceptions of bioengineered foods vary widely and are influenced heavily by the information—or lack thereof—available. Misinformation abounds, with some sources aggressively condemning all genetically modified foods as hazardous while others assert their complete safety and benefit to society. This dichotomy leaves consumers confused and often mistrusting bioengineered foods, regardless of the risk. The debate is further complicated by the marketing practices of food companies that may choose non-GMO labels to imply that their products are somehow healthier or safer despite the lack of scientific consensus supporting these implications. This marketing strategy taps into many consumers' fears and uncertainties about bioengineered foods, driving a market trend that often values perception over scientific reality.

As we continue to explore the complexities of bioengineered foods and their regulation, it becomes clear that there is a critical need for more precise information and robust regulatory standards. Consumers deserve to know whether the food they buy is bioengineered and what that means for their health and the environment. Only then can they make informed choices based on facts rather than fear.

Substances such as trans fats, artificial dyes, and certain preservatives have been introduced into the food industry. These additives are often embraced for their ability to enhance appearance, prolong shelf life, or modify flavor. However, their safety has frequently been questioned, leading to significant health concerns and public debates.

Trans fats, for instance, were once a staple in processed foods, valued for their ability to maintain stability and extend product shelf life. However, scientific studies over the years have consistently linked trans fats to an increased risk of heart disease, prompting a slow but eventual shift in regulatory policies. The timeline of trans fats reveals a troubling delay between initial scientific understanding of their risks and regulatory action. It was not until the early 2000s that significant steps were taken to reduce and eventually ban trans fats from the food supply—a change driven mainly by persistent consumer advocacy and mounting public health data.

Artificial dyes present another chapter concerning the history of food additives. These color additives were initially derived from coal tar—and later from petroleum—and designed to make food more visually appealing. However, research has linked certain synthetic dyes to behavioral issues in children and potential carcinogenic effects. Case studies, such as the infamous "Southampton Six" study, which examined the impact of six common food dyes on children's behavior, have spurred debates and led some countries to increase regulations. However, despite these concerns, many artificial dyes remain approved by regulatory agencies, often defended by arguments pointing to the lack of definitive causal links or the purportedly low levels of exposure in typical diets.

Public outcry has been pivotal in addressing the risks associated with toxic food additives. Armed with research and public support, consumer advocacy groups have often been the catalysts for change, pushing regulatory bodies to reconsider and sometimes reverse their stances on the safety of certain additives. The removal of brominated vegetable oil (BVO) from soft drinks showcases the power of consumer influence. Originally used to stabilize citrus-flavored soft drinks, BVO was the focus of extensive consumer petitions due to health concerns over its link to organ damage and other serious health effects. The response from major beverage companies, phasing out BVO, underscored a growing trend where consumer safety and public perception significantly impact corporate practices and regulatory standards.

Despite these successes, the ongoing use of questionable additives exemplifies persistent issues in food safety oversight. For instance, butylated hydroxyanisole (BHA) and butylated hydroxytoluene (BHT), commonly used as preservatives in foods like cereals and snack items, continue to be permitted despite being flagged for potential carcinogenic effects. The rationale for their continued approval often hinges on regulatory perspectives that deem current exposure levels as safe based on existing studies. This stance, however, ignites debates about the adequacy of existing research and whether it genuinely captures the long-term effects of cumulative and combined exposures to such additives.

As we scrutinize the history of these substances in our food, it becomes increasingly clear that ensuring food safety is not just about reacting to proven harm but anticipating potential risks. It involves a dynamic and proactive approach to regulation that prioritizes public health over industrial convenience. The push for safer food additives is continuous, reflecting growing scientific insight and evolving public health priorities. In this ongoing dialogue between consumers, industry, and regulators, the ultimate goal remains clear: a food system that upholds the highest safety and transparency standards, ensuring that the foods we enjoy are delicious and unequivocally safe for consumption.

NOTES

CHAPTER 2
THE SCIENCE OF SECRECY

Navigating the labyrinth of food labels might seem straightforward, but it is fraught with subtle manipulations and legislative loopholes that can mislead even the most diligent consumers. This chapter peels back the layers of these deceptions, revealing how the food industry uses nuanced language and regulatory gaps to obscure the truth about what is in our foods. As you explore the intricacies of food labeling, you will uncover the clever tricks and loopholes that manufacturers employ to ensure their products appear healthier or more appealing than they genuinely are. It is a game of perception at the cost of consumer health and informed choices.

2.1 Understanding Food Label Tricks and Loopholes

The serving size is one of the most immediate discrepancies on a food label. Often, these figures appear reasonable at first glance but need to align with the typical amounts people consume in one sitting. For instance, a small bag of chips might be listed as containing three servings, though most would drink it in a single session. This manipulation serves a dual purpose: it makes the product's nutritional content, such as calorie and sugar quantities, seem less daunting. Thus, what you perceive as a harmless snack could, in reality, provide a substantial portion of your daily caloric intake, leading to unintentional overconsumption of unhealthy ingredients.

Vague Ingredient Descriptions

Delving deeper into the ingredient list, terms like "natural flavors" and "spices" frequently populate these sections, providing no clarity on what they entail. These terms are legally permitted gateways for food manufacturers to include a variety of substances without disclosing their specific nature. "Natural flavors" could encompass anything from plant-derived essences to chemicals extracted from natural sources, and "spices" could mean a mix of standard kitchen spices or proprietary concoctions. This lack of specificity prevents you from knowing what you are consuming. It hides potential allergens or substances that could harm your health in the guise of innocuous-sounding ingredients.

Health Claim Manipulations

Further complicating are the health claims plastered across food packaging, crafted to catch your eye and convince you of the product's benefits. Phrases like "low fat" and "supports heart health" can be misleading. A product labeled "low fat" might indeed have reduced fat content but could be loaded with sugars to compensate for lost flavor, which is equally, if not more, harmful in excess. Similarly, a claim such as "supports heart health" does not necessarily mean the product benefits everyone; it could simply contain one ingredient known to benefit heart health, among others that might not. These claims exploit regulatory standards that allow them to be used based on the presence of one or two beneficial ingredients, distracting from the less healthy aspects of the product.

Legislative Loopholes

The effectiveness of food labeling is further undermined by legislative loopholes that allow these practices to persist. Current regulations often only require listing certain ingredients or processes if they reach a predefined threshold, meaning trace amounts of potentially harmful chemicals or additives can be legally omitted from labels. Additionally, the lobbying efforts of powerful food industry groups have led to regulations that favor industry interests over consumer protection. For instance, the pushback against stricter labeling laws means that companies can continue using misleading health claims without significant repercussions, maintaining a status quo prioritizing corporate profits over public health.

As you navigate the supermarket aisles, armed with this knowledge, you become more than a consumer; you are an informed citizen, capable of seeing beyond the facade of food labeling. This understanding empowers you to make choices that align more closely with your health and ethical standards, challenging the industry's norm by demanding greater transparency and integrity in food labeling.

Interactive Element: Label Reading Exercise

To help solidify your understanding of deceptive labeling practices, engage in this simple exercise next time you shop:

- Choose any three products with health claims on their packaging.
- Read the nutritional information and ingredient list closely, noting any discrepancies between the health claims and the contents.
- Reflect on how the product's marketing aligns with its nutritional reality and consider whether it would still be your choice now that you have a deeper understanding of labeling practices.

This exercise is about becoming a smarter shopper and reclaiming your right to know what is in your food.

2.2 The Impact of Preservatives on Long-Term Health

Preservatives are ubiquitous in processed foods, playing critical roles in maintaining product safety and quality by inhibiting the growth of harmful microorganisms and preventing the oxidation that leads to spoilage and color loss. Common preservatives like sodium benzoate, potassium sorbate, and sulfites are explicitly chosen for their effectiveness in extending shelf life and maintaining foods' desired texture and flavor. For instance, sodium benzoate is often used in acidic items like salad dressings and carbonated drinks to prevent mold and yeast growth. At the same time, sulfites preserve the bright color of dried fruits and vegetables, ensuring they remain appealing to consumers.

While these chemicals are invaluable to food preservation, research has begun to draw concerning correlations between long-term consumption of certain preservatives and various health issues. Sodium benzoate, when combined with ascorbic acid (vitamin C), can form benzene, a known carcinogen. Studies have shown that this reaction is more likely to occur in products stored in warm conditions, raising serious questions about the safety of preservatives under less-than-ideal storage conditions. Additionally, sulfites, used to prevent discoloration in dried fruits, have been linked to severe allergic reactions and asthma attacks in sensitive individuals. The digestive disturbances associated with various preservatives are also well-documented, with some compounds altering the gut microbiota, potentially leading to conditions like irritable bowel syndrome and decreased nutrient absorption.

The regulatory standards governing the use of food preservatives are ostensibly designed to keep these additives within safe limits; however, the methods and models used for safety assessments often do not reflect real-world eating habits. For example, the cumulative effects of multiple preservatives consumed over time are rarely considered in regulatory assessments, which typically evaluate the safety of one chemical at a time. This oversight can lead to safety thresholds that, while technically accurate under test conditions, do not account for the cocktail of chemicals consumed daily by the average person. Moreover, the reliance on data from animal studies, which may only sometimes accurately predict human responses, further complicates the safety assurance, raising the specter of unforeseen health impacts from long-term exposure to these chemicals.

In the face of these concerns, there is a growing interest in natural or less harmful alternatives to traditional synthetic preservatives. Substances such as rosemary extract, vitamin E, and fermented products like cultured dextrose offer promising preservative functions without the adverse effects associated with their artificial counterparts. These natural alternatives leverage their antioxidant and antimicrobial properties to preserve food. For instance, rosemary extract prevents spoilage and offers anti-inflammatory benefits, making it a dual-purpose additive that enhances food products' shelf life and nutritional profile.

Despite their benefits, adopting natural preservatives in the food industry could be faster. This hesitancy can be attributed to several factors, including the higher cost of natural substances compared to synthetic chemicals and the potential for variability in their preservative effectiveness, which can depend heavily on the source and processing method of the natural compounds. Additionally, natural preservatives often require larger quantities to achieve the same level of preservation as synthetic alternatives, potentially impacting the flavor and texture of the final product.

Understanding the complexities and consequences of food preservatives is crucial for making informed dietary choices. As consumers, increasing our awareness of these substances and their long-term health impacts enables us to advocate for safer, more natural alternatives in our foods, pushing the industry toward practices that prioritize consumer health over shelf stability and cost-effectiveness. This shift could improve public health outcomes and enhance the overall quality and safety of the food supply, aligning it more closely with the growing consumer demand for transparency and natural ingredients in food production.

2.3 How Food Colorings Alter Our Biology

In the vibrant aisles of our supermarkets, the colors of food products play a crucial role in consumer choices. The visual appeal of food is enhanced significantly by the use of food colorings, which are divided into two main categories: artificial and natural.

Artificial colorings, derived from petroleum and other chemicals, are widely used because of their stability and consistency in color. These include colors like Red 40 and Blue 1, commonly found in candies, beverages, and processed foods. On the other hand, natural colorings come from plant, mineral, or animal sources, such as beet extract or annatto. They are used to achieve a similar visual effect with a label that appeals to health-conscious consumers.

The impact of artificial colorings on human biology has been a subject of scientific study and public concern for decades. Research has indicated that certain synthetic dyes can lead to behavioral changes, particularly in children. Studies such as those leading to the Southampton Study in the U.K. have suggested a link between artificial color consumption and increased hyperactivity in some children. Moreover, there are ongoing investigations into the potential for some synthetic colors to cause allergic reactions and their possible carcinogenic properties. For instance, Red 3, a dye once common in candies and desserts, has been acknowledged by the FDA as a potential carcinogen yet remains used in some food products.

The regulatory landscape regarding food colorings is fraught with controversy. In the United States, the FDA oversees the approval and regulation of food dyes, using a certification process that assesses batch purity and safety. However, critics argue that the FDA's reliance on industry-funded studies may lead to conflicts of interest and less stringent safety evaluations. Internationally, the standards can be even more varied. For example, the European Union has taken more proactive measures by requiring labels to warn of potential adverse effects on children's behavior and activity from specific artificial colors, a move not mirrored by U.S. regulations.

Therefore, consumer awareness and making informed choices are more critical than ever. To navigate this colorful landscape safely, consumers must educate themselves about different food dyes and their potential effects. Reading labels carefully is key—looking not just for the terms "artificial colors" but also for specific names like "tartrazine" (Yellow 5) or Alluraa red" (Red 40). Additionally, for those particularly concerned about the impacts of synthetic dyes, opting for products labeled as containing natural colors or free from artificial colors can be a safer choice. Increasing consumer demand for transparency and safety can also drive more companies to consider natural alternatives or to remove unnecessary colorings altogether.

As we close this exploration of how food colorings influence our biology and the significant role regulations and consumer choices play in this dynamic, it becomes apparent that the vibrancy of food colorings extends far beyond mere aesthetic appeal. The implications for health and well-being are profound, highlighting the necessity for ongoing scrutiny, better regulatory frameworks, and a well-informed public. As we transition into the next chapter, we will delve deeper into the additives and processes that change how our food looks and tastes and affect our health in the long term. The following discussion will continue to unravel the complexities of our food system, aiming to equip you with the knowledge to make choices that are not just good but good for you.

CHAPTER 3
MAJOR CULPRITS

In the vast landscape of our daily diets, lurking amidst the familiar comfort of baked goods and the crispy allure of fried snacks, trans fats have positioned themselves as a stealthy danger to public health. Often hidden under the guise of "partially hydrogenated oils" in ingredient lists, trans fats are not just another component of your food. They are a significant risk factor for heart disease, cloaked in the innocuous form of everyday eats from your local grocery store to the fast food corner downtown.

3.1 Trans Fats and Heart Disease: A Hidden Epidemic

Trans fats are a type of unsaturated fat that has been chemically altered through a process known as hydrogenation. This method transforms liquid vegetable oils into a solid form, making them more stable and less likely to spoil. This technique has been celebrated for its ability to prolong shelf life and enhance the flavor profile of foods. Found predominantly in baked goods, fried foods, and processed snacks, trans fats have been a staple in culinary production, valued for their economic and practical benefits in food preservation and texture.

However, the cost to public health is steep. Extensive research has established a robust link between trans fats and an increased risk of heart disease. These fats contribute to the buildup of plaque within the arteries, a condition known as atherosclerosis, which can lead to blockages, heart attacks, or strokes. One pivotal study published in the "New England Journal of Medicine" highlighted that for each increment of 2% in the intake of trans fat, the risk of coronary heart disease increases by a staggering 23%. This statistic is not just a number; it is a ticking time bomb regarding public health implications, signaling an urgent need for regulatory measures and consumer awareness.

Recognizing the gravity of this issue, the FDA has taken steps to eliminate trans fats from the American food supply. In 2015, the FDA released a determination stating that partially hydrogenated oils are not "Generally Recognized as Safe" for use in human food. This decision set the stage for a three-year timeline to phase out trans fats by June 2018 completely. However, the road to a trans-fat-free nation is fraught with challenges. Specific sectors of the food industry have resisted these changes, citing cost concerns and the technological challenges of finding alternatives that offer the same benefits as trans fats.

Moreover, enforcement and compliance with these regulations present their hurdles. While significant food manufacturers may have the resources to reformulate their products, smaller companies often need help with the financial and logistical aspects of such a transition. Additionally, loopholes exist that allow foods containing less than 0.5 grams of trans fat per serving to be labeled as "0 grams of trans fat," potentially misleading consumers who are trying to avoid these fats altogether.

The public health implications of trans fat consumption extend beyond the United States' borders. In countries where trans fats are not regulated, the risks associated with their consumption remain a significant public health concern. For consumers in these regions, the battle against trans fats is even more daunting due to a lack of awareness and regulatory protection.

For those seeking to navigate this challenging landscape, knowledge is power. Understanding how to read food labels correctly is crucial. Look for terms like "partially hydrogenated oil" in the ingredient list—not just the nutritional facts panel—to ensure you are avoiding trans fats. Opting for whole, unprocessed foods when possible can also significantly reduce the risk of trans fat consumption.

Visual Element: Understanding Food Labels

Here is a visual guide to deciphering food labels to help you recognize hidden trans fats. This infographic illustrates how to spot partially hydrogenated oils in the ingredients list and highlights the importance of checking serving sizes to understand the proper amount of trans fats in your food.

By arming yourself with knowledge and making informed dietary choices, you take an active stand not just for your health but for the well-being of your community. Each decision to avoid trans fats is a step towards a healthier heart and a clearer conscience, knowing that you are no longer a passive participant in this hidden epidemic but an informed consumer ready to make choices that foster longevity and vitality.

3.2 High Fructose Corn Syrup and Obesity

High Fructose Corn Syrup (HFCS) has quietly woven itself into the fabric of our daily consumption, predominantly because it is an inexpensive and highly effective sweetener. Derived from corn, HFCS is extensively used in many products such as sodas, sweets, and processed foods. Its widespread use can largely be attributed to its potent sweetness, which surpasses regular table sugar, and its ability to enhance the shelf life of products it is added to. The economical production cost of HFCS also plays a significant role in its prevalence, making it a go-to option for many food manufacturers seeking cost-effective sweetness solutions.

The link between high intake of HFCS and rising obesity rates, particularly in the United States, is supported by a growing body of research. Studies indicate that the consumption of HFCS can lead to significant weight gain and contribute to the development of metabolic syndrome, a cluster of conditions that increase the risk of heart disease, stroke, and diabetes. The biological mechanisms by which HFCS affects the body are multifaceted. Like other sugars, HFCS primarily influences insulin levels and disrupts normal metabolic processes. Unlike glucose, which the body utilizes for energy, fructose is processed in the liver and converted into fat. This conversion can increase fat accumulation, particularly visceral fat around the abdominal area, a significant risk factor for metabolic diseases. Additionally, HFCS consumption does not trigger the body's mechanisms for satiety, as natural sugars might, often leading to overeating.

The corn industry, however, presents a different narrative. Advocates for HFCS argue that it is no different from other forms of sugar and that all sweeteners are safe when consumed in moderation. They point to studies funded by industry stakeholders that suggest HFCS does not directly cause obesity; instead, it is excessive calorie intake, regardless of the source, that leads to weight gain. This stance has sparked a significant scientific debate, with public health advocates pointing to the distinct metabolic effects of fructose on the body as a clear differentiator from other sugars. The debate continues as more studies are undertaken, but the cautionary stance from many health experts remains firm, advocating for reduced consumption of HFCS amid rising health concerns.

The first step for consumers navigating this complex issue is increasing awareness and understanding of HFCS in their diet. Reading food labels is crucial, as HFCS is often hidden under names such as maize, glucose-fructose, and corn sugar. Avoiding products that list any form of high fructose corn syrup among the first few ingredients can significantly reduce one's HFCS intake. Moreover, opting for foods labeled as having no added sugars or choosing products sweetened with natural sugars can also be beneficial. Alternative sweeteners such as honey, agave nectar, or maple syrup, while still sugars, do not contain fructose in the same proportions as HFCS and are metabolized differently by the body.

In addition to making informed choices about sweeteners, integrating whole, unprocessed foods into one's diet can drastically reduce inadvertent HFCS consumption. Fruits, vegetables, whole grains, and lean proteins do not naturally contain HFCS and provide many nutrients without the empty calories associated with many processed foods. This shift towards more natural food sources helps manage weight and supports overall health by providing a balanced spectrum of nutrients for optimal body function.

Navigating the sweetened landscapes of our grocery aisles with an informed perspective allows you to make choices that align better with your health goals. While HFCS remains a topic of contention within nutritional science and public health arenas, the power to influence your health through diet remains firmly in your hands. Making thoughtful food selections based on a comprehensive understanding and sound science is paramount in clarifying the potential pitfalls of high fructose corn syrup and similar additives pervasive in the modern diet.

3.3 MSG and Its Neurological Effects

Monosodium Glutamate, commonly known as MSG, is a flavor enhancer that's been a staple in culinary practices for decades. It is particularly prominent in Asian cuisine and an array of processed foods globally. This sodium salt of the amino acid glutamic acid enhances savory flavors, significantly making dishes appealing without altering their original taste profiles. Its role extends beyond taste enhancement; MSG also helps reduce sodium intake by providing a more complex flavor than salt, allowing less sodium in recipes.

Despite its widespread use and FDA approval, MSG has been surrounded by controversy, particularly concerning its potential neurological effects. The debate centers on what is often referred to as "Chinese Restaurant Syndrome," where consumers report symptoms such as headaches, fatigue, muscle tightness, and numbness after consuming foods containing MSG. These symptoms collectively form what is referred to as the "MSG symptom complex." Scientific studies into these claims have produced mixed results. While some individuals report acute reactions to foods containing MSG, double-masked tests often find no consistent effects when subjects are unaware of MSG's presence in their food. This discrepancy suggests a possible psychological component to the symptoms, though it does not entirely rule out sensitivity in specific individuals.

The FDA currently classifies MSG as Generally Recognized as Safe (GRAS), aligning with international food safety authorities like the World Health Organization and the United Nations Food and Agriculture Organization, supporting its safety. However, consumer fears persist, fueled by early studies and anecdotal reports of adverse reactions. This has led to a market response where many products now prominently label themselves as "No MSG" or "MSG-free" to alleviate consumer concerns and capture a market segment that avoids this additive.

For those who believe they are sensitive to MSG, navigating the grocery aisles can be challenging. It is often hidden in food labels under different names such as hydrolyzed vegetable protein, autolyzed yeast, or even simply terms like 'stock' or 'broth.' Learning these alternate names and reading labels meticulously is crucial to avoid MSG. Options abound for those seeking to replicate MSG's umami effect in cooking without using the substance. Ingredients like tomatoes, mushrooms, and aged cheeses naturally contain high levels of glutamates, providing a similar flavor-enhancing impact. Additionally, fermented products such as soy sauce or fish sauce can contribute a rich umami profile to dishes without adding MSG.

Understanding MSG's role and the scientific landscape empowers you to make informed dietary choices based on your health considerations and preferences. While MSG remains a safe and popular ingredient globally, whether to consume it rests with each individual, underscoring the importance of transparency and education in food labeling and consumer awareness. As we continue to explore additives in our food supply, it becomes increasingly clear that knowledge is not just power but a fundamental component of dietary autonomy. Knowing what is in your food and its effects allows for empowered decisions aligning with personal health philosophies and nutritional needs.

CHAPTER 4
REGULATORY FAILURES

In the intricate dance of food safety and consumer protection, a disheartening rhythm emerges when the music of corporate interests plays louder than the essential needs of public health. This chapter peels back the curtain on a troubling performance where large food corporations, wielding significant influence, choreograph the interests that play louder than the essential needs of public health. This chapter peels back the curtain on a troubling performance where large food corporations, wielding significant influence, choreograph the regulatory frameworks to their tune, often sidelining the health concerns of millions. As you explore regulatory dynamics, you are not just a spectator but a concerned citizen, equipped to understand and question the integrity of the systems meant to protect us.

4.1 The Influence of Big Food Companies on FDA Policies

Corporate Power in Regulatory Frameworks

The sway of large food corporations over the FDA's policies is not just a matter of lobbying prowess but a structural issue that challenges the foundation of regulatory integrity. With their deep pockets and extensive resources, these companies can often disproportionately influence the regulatory processes governing food safety. This influence is not wielded blatantly but subtly through the strategic placement of industry veterans within the FDA and through immense lobbying efforts that shape the legislative landscape.

For instance, it is not uncommon to find that critical positions within the FDA are filled by individuals who have spent a significant part of their careers in the food industry. Often justified by these individuals' expertise, these appointments raise questions about potential conflicts of interest. The revolving door between the FDA and big food companies means that policies and regulations are sometimes seen through the lens of corporate interests rather than through an unbiased, public health-focused perspective.

Examples of Regulatory Capture

Regulatory capture, where regulatory agencies are manipulated to benefit specific industries rather than the public, manifests distinctly in food safety. One glaring example is the delayed enforcement of food safety violations, where regulatory actions are postponed or weakened, often due to pressure from industry groups. A case in point involves a major food company that continued distributing salmonella-tainted peanut butter, contributing to a widespread outbreak while compliance checks were sporadically enforced.

Another instance of diluted regulatory standards is the labeling requirements for genetically modified organisms (GMOs). Despite public demand for transparent labeling, the FDA's guidelines have remained surprisingly lenient, allowing companies to use vague terms or omit detailed information, thus depriving consumers of their right to make informed choices about their food.

Impact on Public Health Policy

The overarching impact of such corporate influence is a regulatory environment that sometimes compromises safety standards, posing risks to public health. These compromises can lead to the approval of additives and processes that may not have undergone rigorous safety testing under the guise of innovation or economic benefit. The consequence is public exposure to potential health hazards, a scenario that more stringent regulatory practices could mitigate.

The influence of big food companies also stifles the innovation of healthier alternatives. Smaller companies that might want to introduce safer, more sustainable products often find themselves outmatched in a landscape where regulatory hurdles are tailored to the capacities of large corporations, discouraging competition and innovation that could benefit public health.

Calls for Reform

In light of these challenges, there is a growing call for reforms to curtail corporate influence in the FDA. Proposed reforms focus on enhancing transparency in the regulatory process, ensuring that the appointment of crucial FDA positions is free from industry conflicts, and strengthening the enforcement of existing food safety laws. For example, initiatives that require public disclosure of all meetings between FDA officials and industry representatives could shed light on the decision-making process, fostering a culture of accountability.

Moreover, there is a push for the FDA to adopt more rigorous, science-based standards to approve food additives and processes, with an increased emphasis on long-term health impacts. These changes, however, require legislative action, public support, and advocacy, underscoring the role of informed citizens in shaping policies that prioritize public health over corporate profits.

Visual Element: The FDA and Big Food - A Complex Web

An infographic is presented here to illustrate further the complex relationship between the FDA and large food corporations. This visual aid maps out the various connections, highlighting the flow of personnel between the FDA and food companies, major lobbying expenditures, and critical legislative battles that have shaped current food safety regulations. This infographic serves as a tool to visualize the depth and breadth of corporate influence, providing a clearer understanding of the stakes involved in regulatory reforms.

As you digest the insights from this section, remember that the dynamics of regulatory frameworks are not just abstract policies but decisions that affect what ends on your plate and, ultimately, your health. The call for reform is not just about changing policies but about safeguarding the trust placed in those who have the power to shape our food environment.

4.2 Case Study: The Aspartame Approval Controversy

It was dissipated even decades after its initial approval. The journey of aspartame from laboratory discovery to a global diet staple is fraught with controversy, marked by scientific disputes and regulatory challenges that provide a revealing glimpse into the FDA's approval processes. Initially discovered in 1965 by a chemist working on ulcer treatment, aspartame's potential as a sweetener was quickly recognized. However, its path to FDA approval, which came in 1981, was simple.

Several vital studies and decisions punctuate the timeline of aspartame's approval. In 1974, the FDA first approved aspartame for limited use in dry goods, but this decision was soon clouded by concerns about its safety, spurred by data suggesting potential health risks. This led to a stay of approval and a request for further investigations. The controversy deepened in the late 1970s when studies by independent researchers suggested that aspartame might cause brain tumors in rats—a finding that led to one of the first major tests of the FDA's commitment to consumer safety over industry interests. Despite these concerns, in 1981, under a new administration keen on reducing regulatory burdens, aspartame was approved for use in dry foods and tabletop use, followed by further approval in 1983 for use in carbonated beverages.

Unusual conflicts of interest and allegations of manipulated research results marked the scientific debates accompanying these regulatory milestones. For instance, early studies conducted by the manufacturer of aspartame were later criticized for their methodological flaws and the alleged manipulation of data to meet safety thresholds. Critics pointed out instances where adverse effects observed in test animals were reportedly omitted from final results or explained away as anomalies. These allegations were severe enough to prompt investigations by the FDA and the U.S. Congress, revealing a regulatory process that appeared unusually responsive to industry pressures.

Public and scientific opposition to aspartame was fierce and sustained. Alarmed by reports of potential health risks, consumer groups mounted campaigns urging the FDA to reconsider its approval. Prominent scientists and researchers also voiced concerns, highlighting discrepancies and deficiencies in the safety data. However, the FDA's handling of these objections was often perceived as dismissive. The agency relied heavily on industry-funded research to counter independent studies suggesting risks, and it moved forward with approvals even as significant segments of the scientific community called for more rigorous examination.

Reflecting on the long-term implications of the aspartame controversy illuminates its enduring impact on public trust in the FDA and other regulatory bodies. This case exemplifies the potential consequences of allowing economic and political considerations to influence regulatory decisions. It has spurred ongoing debates about the adequacy of the FDA's procedures for ensuring the safety of food additives. It has led to calls for more stringent testing and approval processes. Moreover, the aspartame case has had a chilling effect on how the public perceives new food additives, instigating a more skeptical and questioning attitude towards

FDA approvals and the safety of chemically synthesized food products widely present in the market.

This skepticism has only grown in the digital age, where access to information and the rapid spread of accurate and misleading data can sway public opinion vigorously and quickly. As a result, the legacy of the aspartame approval controversy is not just a historical footnote but a continuing influence on how food safety and regulatory integrity are debated in the public sphere today.

4.3 How Lobbying Shapes Food Safety Standards

In the intricate web of food regulation, the influence of lobbying stands out as a profound force that often steers the direction of food safety policies and standards. Lobbyists representing the food industry's interests deploy various tactics to sway legislative and regulatory frameworks. These tactics range from direct lobbying, where industry representatives meet with lawmakers and regulators to present their case, to the strategic allocation of political contributions that aim to secure the goodwill of influential figures within the legislative arena.

One of the primary mechanisms through which lobbyists exert their influence is through the substantial financial contributions made to political campaigns. These contributions are often viewed as investments toward favorable legislative environments that will benefit the industry in the long run. By funding the campaigns of key lawmakers, the food industry ensures it has access to the corridors of power, where decisions about food safety regulations are made. This access allows lobbyists to present their data, often skewed in favor of the industry, and to argue against rules they deem overly restrictive or financially burdensome.

Lobbyists also craft legislation directly. It is not uncommon for industry lobbyists to draft bills that lawmakers present as their own. These drafts are carefully constructed to protect or promote industry interests under the guise of public health and safety improvements. The complexity of regulatory details often makes it difficult for non-experts, including many lawmakers, to fully understand the implications of the proposed changes, allowing significant alterations to food safety standards to be made quietly and without extensive public scrutiny.

Case Examples of Lobbying Impact

Numerous case studies across the regulatory landscape show the impact of lobbying on food safety standards. For instance, consider the relaxation of specific pesticide regulations, which resulted from intense lobbying by agrochemical companies. Despite strong scientific evidence suggesting that these pesticides posed significant risks to human health and the environment, regulatory agencies were pressured to adjust safety standards, allowing higher residues of these chemicals in food products.

Another illustrative example is the introduction of legislation favoring specific food additives. In one case, a food additive, previously banned due to health concerns, was reinstated after the manufacturer funded a series of studies that purported to show its safety. The lobbying efforts included not only the presentation of this new research but also a concerted campaign to discredit earlier studies showing adverse health effects. The result was a legislative amendment that permitted the additive's use under certain conditions, a move seen as a significant victory for the industry.

Regulatory Response to Lobbying

The response of regulatory bodies like the FDA to lobbying pressures is a complex interplay of science, politics, and public health advocacy. While the FDA is mandated to protect public health by ensuring the safety of the food supply, it operates within a political context that can significantly influence its actions. In some instances, the FDA has taken strong measures to maintain regulatory independence and integrity. This includes rigorous peer review processes for scientific data and increased transparency in decision-making processes.

However, there are also instances where the FDA's responses could have been more robust, often due to intense political and financial pressures. For example, when faced with lobbying by powerful food industry groups, the agency has sometimes delayed the implementation of new safety standards or watered-down regulations to make them less onerous for industry compliance. These compromises can lead to situations where consumer safety is less stringently protected than in an ideal regulatory environment free from external influences.

Strategies for Counteracting Lobbying Influence

Several strategies can be implemented to enhance transparency and accountability to counteract lobbying's influence on food safety regulation. Stricter lobbying laws are essential, including rules that require full disclosure of lobbying activities and expenditures. This transparency would help illuminate the extent of industry influence and allow the public and advocacy groups to hold regulators and the industry accountable.

Furthermore, enhancing public access to the decision-making process is crucial. This can be achieved by requiring regulatory agencies to provide more public forums for debate and by making all scientific data and regulatory decisions available online in an easily accessible format. Such measures would enable consumers and independent scientists to scrutinize the data that underpin regulatory decisions, fostering a more democratic and informed regulatory process.

Lastly, strengthening whistleblower protections is vital. These protections ensure that individuals within regulatory agencies or the industry can come forward with information about undue influence or corruption without fear of retaliation. Robust whistleblower protections help maintain the integrity of the regulatory process by ensuring that internal checks and balances can function effectively.

Lobbying and regulation dynamics are central to the discourse on food safety, revealing the tension between corporate interests and the imperative to protect public health. As the chapter closes, remember that these regulatory battles directly influence the quality and safety of your food. The push for stricter regulations, enhanced transparency, and greater accountability is not just bureaucratic wrangling but a critical fight to ensure that food safety standards are driven by science and public health needs, not corporate profits.

CHAPTER 5
GLOBAL COMPARATIONS

As we venture further into the intricate world of food safety, it becomes increasingly clear that the standards and practices governing what ends on our plates vary significantly globally. This chapter turns the spotlight on how different regions tackle the complex issues of food safety, starting with a detailed comparison between European and American approaches. Understanding these differences broadens our perspective and underscores the global challenge of ensuring safe food for all.

5.1 European vs. American Food Safety Standards

The fundamental philosophies guiding food safety regulations in Europe and the United States diverge in significant ways, rooted deeply in cultural and historical contexts that shape each region's approach to regulation. The European Union (E.U.) operates under the precautionary principle, which emphasizes preventive action in the face of uncertainty. If there is a reasonable suspicion that a particular food additive or process could be harmful, the E.U.'s default position is to restrict its use until it is proven safe.

This approach is evident in the stringent standards and assessments that any food product or additive must undergo before it is allowed on the market.

Conversely, the United States often requires concrete proof of harm before taking regulatory action. This reactive approach means that substances may be permitted unless significant evidence demonstrates their adverse health effects. The U.S. Food and Drug Administration (FDA) emphasizes scientific consensus and detailed risk assessments that weigh the benefits against potential harms, sometimes resulting in the approval of banned or restricted substances in Europe.

Impact on Public Health

The differing regulatory philosophies have tangible impacts on public health outcomes. Europe's precautionary approach aims to minimize risk, leading to safer food standards and lower incidences of food-related health issues. For instance, the stringent restrictions on genetically modified organisms (GMOs) and pesticides in the E.U. are driven by concerns about long-term health impacts, which remain under scientific debate. Studies have shown that these regulations contribute to lower residue levels of certain harmful pesticides in food, potentially reducing health risks associated with exposure to these chemicals.

In contrast, the U.S. approach can lead to higher exposure to certain additives and practices until proven harmful. For example, the use of growth hormones in cattle, banned in the E.U., is permitted in the U.S. under FDA regulations. Supporters argue that these hormones are safe based on existing studies, but in Europe, concerns about potential health risks, including cancer and developmental issues, have led to a total ban.

Consumer Protection Policies

The E.U.'s stringent labeling requirements reflect its commitment to consumer rights and transparency. Food packages must display mandatory GMO labeling and detailed allergen information, empowering consumers to make informed decisions based on their health needs and ethical concerns. This level of transparency is designed to foster trust and safety in the food supply.

The U.S., however, has more lenient labeling standards, particularly concerning GMOs. Voluntary labeling initiatives are often the norm, and while there are federal guidelines, they do not mandate the same level of detail found in E.U. regulations. This disparity in labeling standards reflects broader differences in how consumer rights are prioritized and can lead to confusion and a lack of trust among American consumers who seek to understand more about their food choices.

Case Studies of Cross-Contamination and Recall Responses

Cross-contamination in food production and the subsequent recall responses also highlight differences between the two regions. In Europe, the rapid alert system for food and feed (RASFF) enables quick information sharing about potential food safety threats among E.U. countries, leading to swift, coordinated responses. This system has effectively removed contaminated food from the market, minimizing public health risks.

In contrast, the U.S. has faced criticism for slower recall responses, partly due to the fragmented nature of its food safety system, which is split among various federal agencies. The recall of contaminated peanut butter in 2009, which took months and affected hundreds of products, is a case in point. Delays in communication and coordination among agencies, manufacturers, and retailers contributed to the slow response, exacerbating the impact on public health.

As we navigate these comparative insights, it becomes evident that while no system is perfect, the differences in regulatory approaches, public health impacts, consumer protection policies, and crisis responses between Europe and the United States offer critical lessons. These lessons inform our understanding of global food safety challenges and highlight opportunities for adopting best practices that could lead to more effective and harmonized food safety standards worldwide. By examining these global disparities, we equip ourselves with the knowledge to advocate for improvements in our respective systems, pushing for a safer, more transparent food supply chain that prioritizes consumer health and trust.

5.2 Lessons from Japan: A Case for Stricter Regulations

When considering global food safety standards, Japan stands out for its comprehensive approach, characterized by rigorous risk assessments and exacting compliance requirements for domestic and imported foods. Japanese food safety regulations are underpinned by a robust legislative framework prioritizing public health and consumer protection. For instance, the Food Sanitation Act in Japan sets precise guidelines on the use of additives, stipulates detailed food labeling requirements, and establishes strict protocols for handling and processing food. These regulations are enforced through government oversight and self-regulation among food producers, ensuring a high compliance rate across the industry. Such stringent measures reflect a broader commitment to maintaining the integrity and safety of the food supply chain, from farm to table.

Japan's cultural emphasis on food quality and safety further reinforces these regulatory measures. In Japanese culture, food is not merely sustenance but a form of art and a pivotal component of social life. This deep respect for food influences consumer expectations and industry practices, creating a socio-cultural milieu that inherently values high food safety standards. The general public's high awareness and expectation for safe, high-quality food products drive compliance and encourage continuous improvement within the food industry. This cultural backdrop is crucial in understanding why food safety regulations are not just adhered to out of legal necessity but are embraced as a fundamental aspect of food production and consumption.

Technological innovation is pivotal in Japan's food safety strategy due to modern challenges such as mass production and global supply chains. Japan has been at the forefront of implementing cutting-edge technologies to ensure food safety. Advanced tracing and detection systems are widely used across the food industry to monitor product quality and safety continuously. For example, RFID (Radio et al.) technology is employed to track the journey of food products from the farm to the supermarket shelves. This noticeable level of traceability increases the efficiency of recall processes, if necessary, and enhances consumer confidence in the safety of the food they consume. Furthermore, Japan utilizes sophisticated detection systems that can identify minute traces of contaminants, thereby preventing potential foodborne illnesses before products reach the consumer.

The impact of the Fukushima nuclear disaster in March 2011 had profound implications on Japan's food safety perceptions and regulations. Following the disaster, there was a nationwide concern over radioactive contamination of the food supply, particularly affecting regions near Fukushima. The Japanese government responded by implementing stringent radiation testing protocols for food products from affected areas and significantly lowering the acceptable limits of radioactive isotopes in food. These measures were part of a broader effort to restore public trust and ensure the absolute safety of the food supply. Additionally, the crisis increased transparency in Japan's food safety regulations; information on food testing and safety standards became more accessible to the public, fostering greater engagement and confidence in the regulatory processes.

As we examine Japan's comprehensive and culturally ingrained approach to food safety, it becomes evident that its success is not solely the result of stringent regulations or advanced technology. It is also deeply rooted in a societal commitment to quality and safety, supported by public vigilance and a proactive regulatory framework. The lessons from Japan offer valuable insights into how integrating technology, cultural values, and robust regulatory standards can create an effective food safety system that responds to and prevents crises. As other nations grapple with food safety challenges, Japan's model provides a compelling blueprint for achieving higher standards of food safety and consumer protection in an increasingly complex global food market.

5.3 The Canadian Approach to GMOs

In exploring the global landscape of food safety and regulation, Canada presents a unique approach to managing and regulating genetically modified organisms (GMOs). Unlike the United States, where GMOs are widely accepted and less rigorously labeled, and the European Union maintains stringent restrictions and precise labeling requirements, Canada balances with a regulatory framework supporting biotechnological advancements and ensuring consumer safety and transparency.

Regulatory Framework for GMOs

Canada's approach to GMO regulation is rooted in scientific assessment and public health protection. Governed by the Canadian Food Inspection Agency (CFIA) and Health Canada, the regulatory process for GMOs is comprehensive, evaluating new genetically modified crops for their environmental impact, potential health risks, and overall safety. This process is distinct because it is product-based rather than technology-based, meaning that the nature of the end product, rather than the method of its creation, determines the regulatory pathway. This philosophy posits that genetic changes that produce safe and beneficial outcomes should be encouraged, provided they pass stringent safety assessments.

This regulatory framework ensures that all GMOs intended for sale undergo a rigorous pre-market assessment that can take several years to complete. During this phase, GMOs are scrutinized for potential allergenicity, toxicity, nutritional effects, and environmental impact before they are deemed safe for consumption and released into the market. This thorough process reflects Canada's commitment to maintaining high safety standards while embracing the benefits of biotechnological innovations in agriculture.

Public Perception and Consumer Choice

The Canadian public's attitude towards GMOs is shaped significantly by the transparency and thoroughness of the regulatory process. Surveys show a cautious acceptance of GMO foods, contingent on strict safety assessments and clear labeling. This acceptance is bolstered by mandatory labeling guidelines that require clear indications of bioengineered ingredients, allowing consumers to make informed choices based on personal preferences and health considerations.

The demand for non-GMO products has also increased in Canada, reflecting a global trend towards organic and naturally sourced foods. This consumer preference has encouraged a robust market for non-GMO and organic products, with many Canadian companies highlighting these attributes to meet consumer demand. The availability of clear information and choice empowers Canadian consumers, fostering a proactive engagement with food safety and personal health that aligns with broader public health objectives.

Collaborative Approaches with Industry

Canada's regulatory framework for GMOs is notable for its collaborative approach, involving various stakeholders, including government agencies, biotech companies, food producers, and research institutions. This cooperation is facilitated through public-private partnerships that focus on advancing food safety technologies and practices while supporting innovation in the biotech sector.

For instance, Canadian biotech firms often work closely with government scientists and regulatory bodies during new GMO crops' development and assessment phases. This collaboration ensures that the products are innovative and meet the strict safety standards required for approval. By involving industry stakeholders in the regulatory process, Canada harnesses scientific and technological expertise to enhance food safety outcomes while fostering an environment that encourages innovation and economic growth in the agricultural sector.

Case Study: The Approval Process for a New GMO Crop

A detailed examination of the approval process for a new GMO crop in Canada illustrates regulatory bodies' rigorous and systematic approach. Initially, the developer must provide a detailed dossier containing data on the genetic modifications, intended uses, and comprehensive safety assessments covering potential environmental and health impacts. This submission is reviewed by the CFIA, which assesses ecological safety, and Health Canada, which evaluates health safety.

The review process involves several stages of risk assessment, including laboratory analyses, field tests, and comparison studies with non-modified counterparts to ensure that GMOs do not pose more significant risks than traditional crops. Public consultations also form a crucial part of the process, providing a platform for consumers, scientists, and other stakeholders to voice concerns or support the new product. This inclusive approach ensures that multiple perspectives are considered in the decision-making process, enhancing the transparency and credibility of the regulatory framework.

Following the assessments and consultations, a risk management phase ensues, where any potential risks identified during the review are addressed through specific conditions or restrictions on using the GMO. Only after all safety concerns are satisfactorily resolved does the GMO receive approval for commercialization. This meticulous process underscores Canada's commitment to safety and public trust in its food supply.

As we conclude this exploration into Canada's approach to GMOs, it becomes clear that the country's regulatory practices reflect a nuanced understanding of the complexities associated with genetically modified foods. By prioritizing safety, transparency, and collaboration, Canada protects public health and supports innovation in food technology, setting a commendable standard in the global discourse on GMO regulation. This balanced approach is a valuable model for other nations navigating the challenges and opportunities of biotechnological agriculture advancements.

As this chapter closes, we move forward with a deeper appreciation of the diverse strategies employed by different countries to ensure food safety. Each system offers lessons and insights contributing to our collective understanding of how best to protect and nourish populations worldwide, guiding us toward safer, more sustainable food practices for future generations.

CHAPTER 6
CONSUMER IMPACT

Not just in what we eat but in how our food is prepared and preserved. Once reliant on fresh, locally sourced ingredients, we have shifted towards a landscape dominated by processed foods, a change driven by the demands of modern life for convenience and longer shelf life. But at what cost? This chapter delves deep into the direct correlation between our modern dietary habits and the alarming rise in chronic diseases, shedding light on the ingredients silently wreaking havoc on our health. Here, you will find statistics, actionable knowledge, and public health strategies to reverse these trends and foster a healthier future.

6.1 Linking Diet to Disease: What the Data Shows

Correlation Between Processed Foods and Chronic Diseases

The convenience of processed foods is undeniable. Ready in minutes, available at any hour, and often costing less than fresh produce, these modern staples of the Western diet have seamlessly integrated into our daily routines. However, numerous studies have indicated a stark correlation between the high consumption of processed foods and an increase in chronic diseases such as diabetes, cardiovascular diseases, and certain types of cancer. For instance, a comprehensive study published in the "Journal of Public Health" suggests that diets high in processed meats and refined sugars contribute significantly to the incidence of type 2 diabetes and heart disease. These findings are echoed in global research, which consistently links processed food consumption with poor health outcomes due to high levels of unhealthy fats, sugars, and artificial additives found in these foods.

Specific Ingredients and Their Health Impacts

Diving deeper, it becomes crucial to identify the culprits hidden within processed foods. While extending shelf life and enhancing flavor, certain preservatives and artificial sweeteners have been scientifically proven to pose health risks. For example, sodium nitrate, commonly used to preserve the color and flavor of processed meats, has been linked to an increased risk of colorectal and stomach cancer, as reported in studies by the World Health Organization. Similarly, aspartame, a prevalent artificial sweetener in diet sodas and sugar-free products, has been associated with multiple health issues, including metabolic syndrome and cardiovascular diseases. Understanding these links empowers you to make informed decisions about the foods you consume and avoid.

Dietary Patterns and Public Health Recommendations

Over recent decades, the evolution of dietary patterns reveals a shift towards higher calorie intake and less nutritional diversity. This trend coincides with the rise in obesity and other chronic health conditions. Public health recommendations now emphasize returning to a diet rich in whole foods—fruits, vegetables, whole grains, lean proteins—and minimally processed items. Governmental and health organizations globally advocate for dietary changes that include reducing the intake of trans fats, added sugars, and sodium, all prevalent in processed foods. For instance, the Dietary Guidelines for Americans highlight the need for an increased dietary fiber intake and essential nutrients abundantly available in natural, unprocessed foods.

Role of Dietary Education in Prevention

Education plays a pivotal role in preventing diet-related diseases. Informative public health campaigns and interventions have been successful in several regions, significantly altering community health outcomes. Programs that educate children about nutrition have shown promising results in instilling healthy eating habits early on. For example, the 'Farm to School' initiative in the United States has been instrumental in providing schoolchildren with access to fresh, local produce and education on healthy eating, which studies suggest may lead to improved dietary choices in the long term. By investing in dietary education, we equip individuals with the knowledge to make healthier choices and pave the way for a healthier future generation.

Interactive Element: Personal Diet Audit

To actively engage with this information and apply it to your own life, consider conducting a personal diet audit over the next week. Track what you eat daily and note how many items are processed versus whole foods. At the week's end, reflect on your dietary patterns—what you are doing well and what could be improved. This exercise is not about judgment but about awareness and taking proactive steps toward a healthier lifestyle based on the insights you have gained about the impact of diet on health.

6.2 Allergies and Additives: A Growing Concern

In recent years, we have seen a notable increase in food allergy cases, particularly in industrialized nations where processed foods are a staple. This rise is not merely a statistical anomaly but a significant public health concern that reflects deeper issues within our food systems. Studies indicate that this uptick correlates strongly with the widespread use of certain food additives and chemicals that, while improving shelf life and flavor, may compromise immune system responses, particularly in children whose immune systems are still developing.

The connection between food additives and allergic reactions is well-documented but often overlooked in broader dietary guidelines. For instance, sulfites, commonly used in dried fruits, wines, and condiments to prevent bacterial growth and preserve color, can trigger severe asthma attacks and allergic reactions in susceptible individuals. Similarly, certain artificial food colorings, as well as flavor enhancers like monosodium glutamate (MSG), have been linked to allergic responses, ranging from mild itching to life-threatening anaphylactic reactions. The prevalence of these additives in everyday foods makes it increasingly challenging for individuals with food allergies to navigate their dietary choices without extensive knowledge and vigilance safely.

The impact of food allergies extends far beyond the physical symptoms they induce. For many individuals and families, living with food allergies can significantly alter their quality of life. The constant vigilance required to avoid allergens can lead to anxiety and stress, particularly in social situations where food is involved. Dining out, attending social gatherings, and even simple grocery shopping can become fraught with potential dangers. For parents of children with food allergies, the fear of accidental exposure can be all-consuming, affecting their social interactions and mental health.

Despite these challenges, significant advancements in allergy research and management have offered hope for safer and more manageable living conditions for those affected. Recent developments in diagnostic techniques, for example, have made it possible to identify specific allergens more accurately, allowing for more targeted and effective management strategies. New treatments, such as oral immunotherapy, which involves gradually introducing small amounts of an allergen to build tolerance, are promising to reduce the severity of allergic reactions over time.

Moreover, improvements in food labeling laws have made it easier for consumers to identify potential allergens. Clear labeling of common allergens like nuts, dairy, and shellfish is now mandated in many countries, and there is a growing movement towards more transparent labeling of all additives. This shift helps those with known allergies avoid triggers but also assists the broader population in making more informed choices about the foods they consume, fostering greater awareness of how food affects health and well-being.

As we continue to explore the implications of food additives on health, it is clear that more comprehensive research and stricter regulatory standards are needed to ensure that the additives used in our food supply are safe for everyone. By advocating for more robust safety protocols and supporting research into allergy management, we can work towards a future where food allergies are no longer a growing concern but a well-managed aspect of public health. This effort requires not only the involvement of scientists and policymakers but also the educated participation of consumers in advocating for transparency and safety in the food industry.

6.3 The Rise of Food Sensitivities and Intolerances

In the complex landscape of dietary health, distinguishing between food allergies, sensitivities, and intolerances is crucial for understanding their impacts on our well-being. While food allergies involve an immune system response that can be severe and even life-threatening, food sensitivities and intolerances generally refer to difficulty digesting certain foods, leading to uncomfortable but typically less dangerous symptoms. Food sensitivities often involve an immune response but are usually less severe than allergies. In contrast, food intolerances, like lactose intolerance, arise from the body's inability to properly digest certain substances due to enzyme deficiencies or reactions to food additives without an immune response.

The prevalence of food intolerances has been on the rise, with lactose intolerance and non-celiac gluten sensitivity being among the most common. Lactose intolerance affects a significant portion of the global population, with symptoms including bloating, diarrhea, and abdominal pain after consuming dairy products. This condition stems from the body's reduced ability to produce lactase, the enzyme needed to digest lactose found in dairy. Non-celiac gluten sensitivity, while still not fully understood, presents symptoms similar to celiac disease but without the associated autoimmune response. Individuals report bloating, fatigue, and headaches, which markedly improve when gluten is removed from the diet.

Recent clinical studies and patient reports indicate that certain food additives might exacerbate symptoms of food sensitivities. For instance, monosodium glutamate (MSG), used widely as a flavor enhancer, has been reported to trigger headaches and other symptoms in sensitive individuals. Similarly, artificial sweeteners like aspartame may cause digestive discomfort in some people. These reactions could be due to the additives stimulating or irritating the gut in ways that exacerbate existing sensitivities, highlighting the need for careful consideration of food ingredients among those with dietary sensitivities.

Navigating a diet with food sensitivities involves becoming vigilant about the ingredients in your food. Reading labels becomes paramount; learning to identify the apparent ingredients and those hidden under less recognizable names is critical to avoiding triggers. For example, MSG can also appear on labels as 'hydrolyzed vegetable protein' or 'yeast extract.' Opting for whole and less processed foods can significantly reduce the risk of exposure to problematic additives. When symptoms persist, seeking guidance from health professionals who can offer dietary advice, conducting tests to pinpoint specific sensitivities, and developing a personalized nutritional plan to manage symptoms effectively becomes essential.

Encountering food sensitivities can complicate your relationship with food, transforming everyday eating into a process that requires scrutiny and sometimes difficult choices. However, by arming yourself with knowledge and learning to navigate these challenges, you can maintain your health and enjoyment of food. As we uncover how our bodies react to different dietary components, we hope this knowledge will lead to greater empowerment and well-being for those affected by food sensitivities and intolerances, fostering a more informed and health-conscious society.

Critical Reflections and Next Steps

Exploring food sensitivities and intolerances brings to light the intricate ways our bodies interact with the foods we consume. Understanding the difference between sensitivities, intolerances, and allergies is the first step toward effectively addressing and managing these conditions. The rise in reported cases highlights the need for greater awareness and better food labeling practices, enabling individuals to make informed dietary choices. As we transition into the next chapter, the focus shifts to the industry insiders and experts who will share their insights on navigating the complex terrain of food safety, further clarifying how best to advocate for a safer, healthier food system. This ongoing dialogue is essential for fostering an environment where everyone can make dietary choices that support their health and well-being.

CHAPTER 7
INDUSTRY INSIDERS SPEAK

The journey into the heart of the food industry is often paved with good intentions. Many enter the field of food science driven by a passion to innovate and improve the quality of food products, aspiring to contribute to public health and safety. However, the stark realities of industry pressures and ethical conflicts can reshape these intentions, leading to profound professional and personal dilemmas. In this chapter, we delve deep into the confessions of a food scientist whose career illuminates the darker corners of the food industry, where the drive for profit often overshadows the imperative of health.

7.1 Confessions of a Food Scientist

Journey into the Food Industry

Imagine embarking on a career filled with the promise of improving how people eat and live, only to find that the path needs to be revised with compromises and unsavory practices. This is the reality for many food scientists who start their careers aiming to develop healthier, sustainable food options. Initially, their work is guided by the noble objectives of enhancing nutritional value and ensuring safety. However, as they delve deeper into the industry, they often encounter a pervasive focus on cost reduction, shelf stability, and consumer appeal, which frequently comes at the expense of nutritional integrity and health. The pressure to conform can be overwhelming, forcing food scientists to reassess their roles and the impact of their work on public health.

Exposure to Malpractices

Like any other, the food industry is not immune to ethically questionable or outright harmful practices. Food scientists often witness the intentional misuse of additives meant to enhance flavor or texture or preserve color used excessively to mask inferior quality ingredients. Additionally, manipulating expiry dates raises serious safety concerns, as products that should be pulled from shelves are left to be purchased by unsuspecting consumers. Perhaps most alarmingly, the non-disclosure of potential allergens or toxins, driven by a fear of impacting sales or product image, directly jeopardizes consumer health. While not universal, these practices are sufficiently standard to merit concern and vigilance.

Ethical Dilemmas and Decisions

Confronted with these malpractices, food scientists face significant ethical dilemmas. When research findings or personal observations contradict company policies or reveal harmful consequences, these professionals must make difficult choices. Speaking out against malpractices can mean risking job security, facing professional ostracism, or enduring legal battles. However, silence compromises personal ethics and public safety. The decision to prioritize integrity over personal gain is daunting but crucial. Those who take a stand set a powerful example, shedding light on the need for systemic changes within the industry.

NOTES

CHAPTER 8
IMPACT OF WHISTLEBLOWING

Whistleblowers in the food science arena often endure formidable challenges. The decision to expose wrongdoing can lead to legal repercussions as companies move to protect their interests. Blocklisting within the industry can end careers, making it difficult for these individuals to find employment in their field. Despite these hardships, whistleblowing can have a profound positive impact. It can increase public awareness, spur regulatory investigations, and result in significant policy changes that enhance food safety and protect consumer health. The courage of these individuals often inspires others in the industry to come forward, contributing to a culture of accountability and transparency.

Visual Element: The Whistleblower's Journey

An infographic illustrates the path of a whistleblower in the food industry. It outlines the typical stages, from discovery, decision-making, and speaking out to the aftermath and broader impacts. This visual guide helps contextualize the personal and professional journey of food scientists who dare to expose the truths hidden behind laboratory doors.

In navigating these complex and often hidden aspects of the food industry, you are invited to reflect on the broader implications of these practices. Understanding the challenges industry employees who strive to uphold ethical standards face is crucial for fostering a more transparent and responsible food system. Awareness of these issues enables consumers to make more informed choices and support practices prioritizing health and safety over profit. This knowledge empowers us all to advocate for a food industry that aligns with our values and the well-being of communities worldwideIn an era where glossy media often fails to tell the truth about what's on our plates, it becomes not just relevant but vital. Picture this: an investigative journalist, armed with nothing but a pen and notepad, or perhaps a camera, delving into the labyrinthine dealings of the food industry to unearth truths that can shake the foundations of public health and corporate accountability. This is the frontline of information warfare, where every article and report has the potential to alter the landscape of food safety and consumer awareness significantly.

8.1 Investigative Journalism and Food Safety

Role of Investigative Journalism in Uncovering Food Scandals

Investigative journalism has historically been pivotal in exposing significant food safety issues, often leading to public outcry and catalyzing regulatory changes. Consider the case of the 2013 horse meat scandal in Europe, where investigative reporters revealed that foods advertised as containing beef were made with horse meat. This exposé caused a massive public uproar, leading to tighter food labeling and processing regulations across the European Union. Similarly, in the United States, journalists have uncovered issues ranging from antibiotic misuse in livestock to harmful chemicals in everyday foods, prompting discussions and actions on national food safety policies.

These case studies underscore the power of diligent, on-the-ground reporting in holding large corporations and regulators accountable. By bringing critical information to light, investigative journalists empower consumers with the knowledge to make safer food choices and advocate for changes in the industry.

Challenges Faced by Investigative Journalists

Despite the critical role they play, investigative journalists face numerous obstacles. Legal threats are the most direct challenge, with food corporations often employing teams of lawyers to suppress damaging stories through lawsuits or threats thereof. Access to information is another significant hurdle. Many journalists need help to obtain crucial data, as food companies and government bodies can be opaque, refusing to release relevant files or engage with the media. Moreover, the influence of powerful food industry lobbyists can lead to self-censorship among certain media outlets that fear economic repercussions, such as the loss of advertising revenue from big food companies.

Impact of Exposés on Public Policy and Consumer Behavior

The impact of well-documented investigative pieces extends beyond mere awareness, influencing public policy and consumer behavior. Exposés have led to the enactment of new safety regulations, recalls of dangerous products, and more stringent enforcement of existing laws. They also inform consumer choices, driving demand for healthier, more ethically produced food products. As consumers become more aware of the issues, their purchasing decisions reflect their values more accurately, which pressures companies to adopt better practices.

Ethical Considerations in Reporting

Ethical journalism is the cornerstone upon which public trust is built. Investigative reporters must navigate complex ethical landscapes to report on food safety issues accurately and relatively. They must balance the need for compelling storytelling with the imperative of accuracy, avoiding sensationalism that can mislead rather than inform. Additionally, they must handle sensitive information responsibly, protecting confidential sources while ensuring the public gets access to the information it needs to make informed decisions. The potential impact of their reporting on public health and business reputations places a heavy burden on journalists to get their stories right, adhering strictly to the principles of truth, accuracy, and fairness.

Interactive Element: Critical Media Literacy Exercise

To further engage with the content of this chapter and enhance your understanding of media and food safety, here is an exercise designed to develop your critical media literacy skills. Choose a recent article on food safety and evaluate it based on the following criteria:

- **Source Credibility:** Who reports the story, and what is their reputation?

- **Evidence:** What evidence is presented, and how is it sourced?
- **Balance:** Does the report present multiple viewpoints, and are they fairly represented?
- **Impact:** What is the report's impact on public perception and policy?

This exercise will help you discern the quality and reliability of information you encounter daily, empowering you to make better-informed decisions about the food you consume and the information you share.

As the chapter unfolds, the narratives of dedication, risk, and integrity woven through the world of investigative journalism highlight challenges and the profound impacts of this field on public health, policy, and awareness. Through their relentless pursuit of truth, investigative journalists ensure that the public remains informed and the powerful are held accountable, playing an indispensable role in the ongoing dialogue about food safety and public health.

8.2 Mainstream Media vs. Independent Reporting on Health Issues

Regarding reporting on food safety, the landscape of information dissemination is starkly divided between mainstream media and independent outlets, each wielding its influence but in markedly different manners. With its broad reach, mainstream media often shapes public perception and discourse on health issues. However, it is also subject to influences that can dilute the depth and breadth of its coverage. In contrast, independent media, unshackled by corporate ties, often dives deeper into issues that may escape the radar of their mainstream counterparts.

The coverage of food safety by mainstream media tends to be episodic and reactive, focusing on specific incidents of food recalls or foodborne illness outbreaks as they happen. While effective in alerting the public to immediate risks, this approach often needs a sustained investigation into the underlying systemic issues that permit such crises to recur. Moreover, the breadth of coverage can be influenced by the need to cater to a broad audience or to avoid alienating advertisers, mainly when those advertisers are influential players within the food industry. This conflict of interest can lead to reporting that sometimes skirts around deeper issues or fails to confront the responsible parties, such as large agricultural corporations whose practices might be at the heart of systemic food safety failures.

In contrast, independent media outlets often operate with more freedom to explore and expose the complexities behind food safety breaches. Without the looming pressure of losing advertising dollars from big food companies, these outlets can afford to bite into the malpractices and policies that mainstream media may touch only superficially. For example, independent investigations might track the long-term health implications of repeated pesticide use in farming or the failures in regulatory frameworks that allow risky food products to reach consumers. This depth of coverage informs and empowers consumers and policymakers to demand more stringent safety measures.

The bias in mainstream media can often be traced back to its structural dependencies. Extensive networks and publications depend on advertising revenue to survive, and biting the hand that feeds them can be considered risky. This dependency can lead to a certain degree of self-censorship or to a phenomenon known as 'framin',' where the way a story is told—and what is left out—can subtly align with the interests of the advertisers. For instance, coverage of unhealthy food products might underplay the health risks associated with their consumption or may fail to confront the aggressive marketing tactics used to target vulnerable populations like children.

The advantages of independent media in the realm of food safety are manifold. By prioritizing public interest over profits, these platforms can cover stories that might be deemed too controversial or economically risky by mainstream channels. They are often the first to provide a platform for whistleblowers or to uncover data that corporations would rather keep confidential. The freedom from corporate influence allows independent journalists to craft their narratives purely based on factual evidence and the story's significance to public health rather than on its potential to generate ad revenue.

One compelling instance where independent media has significantly impacted involves the reporting on antibiotic resistance linked to the overuse of antibiotics in livestock. While mainstream media coverage was sporadic and largely reactive, independent outlets provided extensive and ongoing coverage that highlighted the problem of the lack of adequate regulatory measures to address it. This sustained attention helped ignite public and policy debates crucial for driving change in regulatory policies and farming practices.

As you navigate the complex narratives presented by different media outlets on food safety issues, understanding their inherent biases and strengths can significantly enhance your ability to evaluate the information presented critically. It empowers you to seek sources that prioritize depth, accuracy, and independence over sensationalism or profitability. In doing so, you become not just a passive consumer of information but an active participant in the discourse on food safety, equipped to make informed decisions and advocate for changes that enhance the wellbeing of all.

8.3 Social Media's Influence on Food Trends and Misinformation

The digital age has transformed how information is disseminated and consumed, particularly concerning food safety. With their vast reach and immediacy, social media platforms have become powerful tools for raising awareness about food safety issues, mobilizing grassroots movements, and advocating for change. Platforms like Twitter, Facebook, and Instagram allow for rapid information sharing, enabling campaigns that advocate for healthier, safer food practices to gain traction quickly and engage a broad audience. For instance, social media campaigns like #EatLocal and #FoodSafety have helped spread knowledge about the benefits of consuming locally sourced foods and the importance of adhering to safety practices in food preparation.

Despite these positive aspects, social media poses significant challenges, particularly concerning the rapid spread of misinformation. The nature of these platforms often prioritizes sensational or emotionally charged content, regardless of its accuracy. This can lead to the viral spread of misinformation about food safety, where unfounded claims or outright false information gain widespread attention. Misinformation can undermine public health efforts, create unwarranted fears, or lead to unsafe food practices based on incorrect information. The consequences can be severe, detracting from scientifically backed health campaigns and potentially leading to public health risks.

Influencers on social media play a pivotal role in shaping consumer perceptions about food safety and health trends. These individuals, often with large followings, can significantly impact public opinion and behavior with just a single post. When influencers share content about diet trends or highlight specific food products, they can sway massive segments of the population, for better or for worse. However, the credibility of these influencers can vary greatly. While some may base their recommendations on sound nutrition science, others might propagate myths or unverified facts, contributing to confusion and misinformation about food safety.

Several strategies can effectively combat the spread of misinformation on social media. First, education plays a crucial role. By improving public understanding of food safety and nutrition, individuals are better equipped to evaluate the information they encounter online critically. Initiatives like digital literacy programs that teach users to verify sources and check facts can empower consumers to sift through misinformation. Additionally, fact-checking services provided by social media platforms can help flag or remove false information, although the implementation and effectiveness of these services can vary.

Social media platforms monitor and manage shared content, especially concerning public health. Collaborating with health experts and organizations to create verified content can help ensure the information reaching the public is accurate and beneficial. Moreover, platforms can design algorithms that prioritize evidence-based information and demote content flagged as false, although this approach requires careful balance to maintain freedom of expression.

As we navigate the complex interplay of food safety, public health, and digital communication, it becomes clear that social media is a double-edged sword. While offering unprecedented opportunities for advocacy and education, it also challenges us to be vigilant and critical information consumers. As this chapter concludes, we recognize the power of social media to influence food trends and public health discussions. Moving forward, it is crucial that all stakeholders, from individuals to platform developers, work collaboratively to harness this power responsibly, ensuring that the digital discourse on food safety contributes positively to public knowledge and wellbeing.

This exploration of social media's shaping of food safety perceptions underscores the transformative impact of digital platforms on public discourse. As we close this chapter, we are reminded of the dual nature of social media — its ability to both enlighten and mislead. Our collective responsibility is to foster an online environment where truth prevails, and public health is safeguarded. Looking ahead, the next chapter will delve deeper into the legal battles surrounding food safety, where the courts become arenas for defending public health and ensuring. In the intricate dance of food safety and consumer rights, the courtroom has often served as a battleground where the scales of justice weigh heavily against towering food corporations. Here, the voices of everyday consumers, armed with evidence and the law, challenge practices that threaten public health. This chapter delves into the pivotal legal disputes that have not only shaped the landscape of food safety but have also underscored the power of legal recourse in correcting industry wrongs.

CHAPTER 9
LANDMARK LAWSUITS IN FOOD SAFETY

Historical Overview of Impactful Cases

The legal history surrounding food safety is marked by several landmark cases that have catalyzed significant changes in how food is regulated and marketed. One of the earliest examples, the infamous case against a major food company in the late 1990s, involved misleading advertising about the health benefits of a cereal brand. The lawsuit ended in a multi-million-dollar settlement and set a precedent for how health claims could be used in food marketing, leading to tighter regulations by the FDA.

Another pivotal case occurred in the early 2000s when a fast-food giant was sued for misrepresenting its frying oil. The suit claimed that the company did not adequately disclose the use of unhealthy trans fats, which contribute to public health risks like heart disease. The outcome was a substantial shift in the company and the entire fast-food industry, with many chains moving to healthier alternatives and more explicit nutritional labeling.

These cases, among others, have been instrumental in exposing deceptive practices in the food industry, compelling more transparency and accountability from food manufacturers and distributors.

Analysis of Legal Outcomes and Their Implications

The ramifications of these legal battles extend far beyond the courtroom. Each case, especially those favoring consumer rights, has ripple effects throughout the food industry. For instance, settlements often include financial penalties and company commitments to change their labeling practices, ingredient disclosures, and advertising strategies. These changes are not merely cosmetic; they represent shifts in industry standards and practices, influencing how food safety is approached and managed across the board.

Moreover, these legal outcomes often inspire regulatory reforms. Legislators and regulators, spurred by the publicity and outcomes of high-profile cases, may introduce or tighten food safety laws and enforcement measures. This proactive legal environment deters potential malpractice and fosters a culture of compliance and vigilance within the food industry.

Role of Class Action Lawsuits

Class action lawsuits have been particularly effective in addressing widespread public health issues related to food safety. These collective legal actions enable individuals who might otherwise lack the resources to sue large corporations to band together and strengthen their legal stand. An example was a class-action lawsuit against a beverage company accused of using a carcinogenic ingredient. The lawsuit represented thousands of consumers who were concerned about the potential health risks. The settlement led to removing harmful ingredients from the product, a victory for consumer safety on a large scale.

Such lawsuits correct injustices and serve as powerful deterrents, signaling to the food industry that public health cannot be compromised without consequence. The threat of potential lawsuits can lead companies to be more rigorous in their compliance with food safety regulations and more transparent with their product information.

9.1 Recent and Ongoing Legal Battles

The food safety landscape continues to evolve, and so do the legal challenges. One of the current legal battles involves the labeling of GMO products. As consumer demand for transparency grows, several states have seen lawsuits challenging the adequacy of GMO labeling laws. These cases could set new precedents for how genetically modified foods are presented to consumers, potentially leading to stricter labeling requirements.

Another ongoing issue is the legal response to foodborne illness outbreaks. Recent cases against restaurant chains and food producers have highlighted gaps in food safety practices that allowed contaminants to enter the food supply. These lawsuits seek compensation for those affected and push for systemic changes to prevent future outbreaks, such as better tracking of food sources and stricter contamination controls in food processing.

Visual Element: Timeline of Landmark Food Safety Lawsuits

A detailed timeline infographic is included to clarify how legal actions have shaped food safety. This visual element traces key lawsuits from the past decades, highlighting major legal battles, their outcomes, and their impacts on food safety regulations and practices. This timeline serves as a visual testament to the power of legal action in enforcing food safety and protecting public health.

As we explore the dynamic interplay between law, consumer rights, and food safety, it becomes apparent that legal battles are not just about disputes but about upholding transparency, accountability, and public health principles. These legal frameworks and actions reinforce the notion that food safety is a right, not a privilege, ensuring that the food industry remains responsible for the wellbeing of its consumers.

9.2 The Power of the Freedom of Information Act (FOIA)

The Freedom of Information Act (FOIA), enacted in 1966, is a pivotal tool in the quest for governmental transparency, allowing the public unprecedented access to federal agency records. This act empowers you, the citizen, to peel back the curtain on government operations, including those concerning food safety regulations and enforcement actions. Essentially, FOIA is built on the principle that an informed citizenry is essential to the functioning of a democratic society. It helps ensure that government actions aren't shrouded in secrecy and that officials are held accountable.

Through FOIA, individuals and organizations can request documents and data that reveal how food safety policies are developed and implemented. These records include internal communications, scientific data, inspection reports, and regulatory decisions. For instance, FOIA requests have unveiled discrepancies in enforcing food safety laws, where certain violations were ignored or inconsistently handled. They have also exposed cases in which lobbying by powerful food industry groups influenced regulatory decisions, sometimes to the detriment of public health standards. These revelations often lead to public outrage, prompting regulatory reforms or more stringent enforcement of existing laws.

One notable FOIA revelation involved a major poultry producer and uncovered how lapses in regulatory oversight allowed the sale of chicken contaminated with harmful bacteria. This information led to a nationwide health advisory and stricter controls on poultry inspection processes. Another significant use of FOIA revealed the extent of pesticide residues in everyday fruits and vegetables, pushing the Environmental Protection Agency (EPA) towards tighter regulations on pesticide use in agriculture.

Despite its critical role, the path to obtaining information via FOIA needs to be improved. One major hurdle is the frequent delays in processing requests. Agencies often cite a backlog of requests or limited resources as reasons for delays, extending from a few months to several years. This frustrates the requesters and delays disseminating potentially crucial information to the public. Moreover, when agencies do respond, the documents provided can be heavily redacted, with substantial portions blacked out for reasons of confidentiality or security. This can severely limit the usefulness of the information, leaving gaps that can hinder comprehensive understanding and analysis.

Another challenge is the outright denial of requests. Agencies might refuse FOIA requests on various grounds, such as protecting personal privacy, national security, or internal agency matters. While some denials are justified, others are seen as attempts to prevent embarrassing or damaging information from becoming public. Navigating these denials often requires legal expertise and can lead to appeals and lawsuits, complicating what was intended to be a straightforward process of information gathering.

Several reforms could enhance the effectiveness of FOIA. Reducing the exemptions that allow agencies to withhold information would make more documents eligible for public release. Simplifying the request process and improving the digital submission systems could help manage and reduce the backlog of requests. Agencies could also be mandated to proactively publish critical documents, particularly those about public health and safety, which would decrease the need for individual requests. Additionally, increasing funding and resources dedicated to FOIA offices would help improve response times and reduce delays.

As a powerful tool for ensuring governmental transparency, FOIA is indispensable in advocating for safer, healthier food systems. By shedding light on the often opaque food safety regulation processes, FOIA informs and empowers individuals and organizations to advocate for necessary changes and hold authorities accountable. In the ongoing dialogue about food safety, the ability to access information freely remains a cornerstone of effective advocacy and informed public debate.

9.3 Advocacy Groups and Their Role in Legal Reforms

Advocacy groups play a crucial role in the ongoing effort to safeguard public health through better food safety standards. These organizations, driven by a commitment to public wellbeing and consumer rights, often stand as the first defense against lax practices and policies in the food industry. By mobilizing resources, expertise, and public support, these groups work tirelessly to bring about legal and regulatory changes that protect consumers from potential harm caused by unsafe food practices.

One of the leading organizations in this field is the Center for Food Safety (CFS), a national non-profit organization known for its rigorous advocacy for organic standards and against genetically modified foods. CFS aims to protect human health and the environment by curbing harmful food production technologies and promoting organic and sustainable agriculture. Another prominent group, Food & Water Watch, focuses on a broader spectrum of issues, including food, water, and environmental practices, advocating for policies that ensure safe, accessible, and sustainably produced food and water.

The strategies employed by these advocacy groups are both diverse and dynamic. Lobbying remains a core tactic, involving direct interactions with lawmakers to advocate for the passage of stronger food safety laws. For instance, these groups frequently provide scientific data and policy analyses that help shape legislative debates on food safety. Public awareness campaigns are another critical strategy. By educating consumers about the risks associated with specific food practices and the importance of stringent safety standards, these groups build public pressure on policymakers to act. Moreover, many organizations form alliances with other stakeholders, including environmental groups, consumer coalitions, and health professionals, to amplify their voices and extend their influence.

A notable success story is advocacy groups passing the Food Safety Modernization Act (FSMA) in 2011, the most sweeping reform of food safety laws in over 70 years. Organizations like the

CFS were instrumental in highlighting the gaps in the existing framework and advocating for a system that focuses more on preventing food safety problems before they occur. The FSMA now requires more comprehensive preventive controls for food facilities, introduces stricter standards for produce safety, and mandates better oversight of imports.

Despite these successes, advocacy groups face significant challenges. One of the most daunting is the opposition from powerful industry lobbyists who often have deeper pockets and more access to political influencers. These lobbyists work to counteract the efforts of advocacy groups, pushing for less restrictive laws that favor industry profits over consumer health. Moreover, advocacy groups can need more resources to sustain long-term campaigns or respond quickly to emerging issues. The complexity of food safety legislation and the scientific intricacies involved can also pose challenges, requiring groups to maintain high technical expertise.

The ongoing battle for safer food practices is a testament to the resilience and dedication of these advocacy groups. Their efforts challenge inadequate practices and policies and contribute to a broader movement towards a safer, more transparent food system. As we continue to navigate the complexities of food safety, the role of these organizations remains indispensable, advocating for the changes necessary to ensure that the food on our tables is safe to eat. Their relentless pursuit of reform protects public health and reinforces the principles of accountability and transparency in the food industry.

As this chapter on legal battles and the role of advocacy groups concludes, we are reminded of the powerful impact collective action and informed advocacy can have on shaping public policy.

The efforts of advocacy groups, often unsung, play a crucial role in the ongoing dialogue about food safety, pushing boundaries and challenging the status quo. Their work ensures that food safety remains a top priority in public discourse and legislative action, safeguarding public health and consumer rights. As we move forward into the next chapter, we carry with us a deeper understanding of the legal frameworks that govern our food system and the vital role that informed citizens and organized advocacy play in maintaining the integrity and safety of our food supply.

NOTES

CHAPTER 10
ETHICAL CONSIDERATIONS

Imagine standing in a supermarket aisle, scrutinizing the back of a food package, trying to decipher what exactly goes into the products you consume and how they are made. It's not about personal health; it's about making ethical choices in a world where food production significantly impacts society and the environment. The food industry, a potent force in the global economy, holds a profound ethical responsibility towards consumers and the planet. This chapter delves into the moral obligations of food manufacturers, the necessity of transparency, the environmental and social responsibilities they shoulder, and some cautionary tales of ethical breaches that have reshaped public trust.

10.1 Ethical Responsibility of Food Manufacturers

Duty to Protect Public Health

At the heart of the ethical obligations of food manufacturers is the duty to ensure the safety and health of their products. It's a

Fundamental principle that should guide every decision in the food production process. Yet, too often, this fundamental moral obligation is overshadowed by the pursuit of profit. The implications of neglecting this duty stretch far beyond the corporate bottom line—they ripple out to affect the health of millions. Food manufacturers must rigorously test and ensure their products are safe for consumption, adhering to the minimum legal standards and the highest ethical expectations. This responsibility also involves proactively updating practices as new health information becomes available, ensuring consumer safety always takes precedence.

Transparency with Consumers

Transparency is the cornerstone of building trust between consumers and manufacturers. It involves clear, honest communication about how food products are sourced, processed, and labeled. In today's consumer age, people demand to know more than just the nutritional content of their food; they want to understand where it comes from, how it was made, and what it contains. Transparency isn't just about listing ingredients; it's about providing context and why they are used. This openness is essential for building consumer trust and empowering consumers to make informed choices about the foods they eat, aligning with their values and health needs.

Environmental and Social Responsibility

The ethical responsibilities of food manufacturers extend beyond their consumers. They encompass broader stewardship of the planet and its resources. This includes sustainable sourcing of ingredients, minimizing environmental footprints, and ensuring fair labor practices throughout the supply chain. For instance, palm oil is an ingredient in numerous food products, from biscuits to frozen meals. However, its production is often linked to significant environmental degradation, including deforestation and loss of biodiversity. Ethical food manufacturers commit to using sustainably sourced palm oil, demonstrating a commitment to ecological conservation. Similarly, ensuring that workers in all stages of production are treated fairly and work in safe conditions reflects a commitment to social responsibility. These practices are not just ethical imperatives; they are increasingly becoming expectations from consumers who place a premium on sustainability and ethical business practices.

Case Studies of Ethical Breaches

Reflecting on past ethical breaches provides critical lessons for consumers and the industry. One illustrative example is the case of a famous instant noodle brand in India that was found to contain lead levels exceeding permissible limits. The fallout was massive, with widespread public outrage, a temporary ban on the product, and a significant hit to the competition. This case highlighted the health risks of neglecting product safety and the long-term business risks of losing consumer trust. Another example involves a major U.S. poultry producer that faced allegations of falsely advertising their chicken as "humane" and raised" despite" evidence of inhumane treatment at their processing facilities. The backlash from consumers and watchdog organizations was swift and severe, leading to legal battles and a public relations crisis.

WHAT THE FDA & THE GOVERNMENT ARE COVERING UP!

Visual Element: Ethical Practices Checklist

A comprehensive checklist is provided to further engage with ethical responsibility in food manufacturing. This checklist serves as a guide for consumers to assess a commitment to ethical practices. It includes questions about ingredient sourcing, labor practices, environmental impact, and transparency in labeling. This checklist allows consumers to make more informed choices, supporting companies aligning with their ethical standards and encouraging manufacturers to adopt responsible practices.

As we navigate the complexities of ethical considerations in food manufacturing, it becomes clear that the choices made at every step of the food production process have profound implications. From the health of consumers to the health of the planet, the decisions of food manufacturers bear weighty ethical responsibilities. By adhering to safety, transparency, and sustainability principles, the food industry can uphold its moral obligations and build a legacy of trust and integrity with consumers and society.

10.2 Consumer Rights vs. Corporate Security

The tug-of-war between the consumer's knowledge and a corporation's protection of its trade secrets presents a complex ethical battlefield in the modern food industry. On the one hand, as a consumer, you rightfully demand to understand precisely what you are consuming — where the ingredients come from, how they are processed, and whether they contain genetically modified organisms (GMOs) or allergens. On the other hand, food manufacturers argue that revealing detailed recipes and processes could compromise their competitive edge, equating it to giving away the secret sauce—literally. Balancing these interests requires a

Ethical Considerations

nuanced approach, blending legal frameworks with ethical considerations to safeguard public health and private enterprise.

The balance between transparency and confidentiality is not just a theoretical debate but has real-world implications for public health. Consider the scenario where a food product causes adverse health reactions among consumers. Suppose the company had not disclosed all ingredients, citing trade secrecy. In that case, medical professionals might have struggled to pinpoint the cause, delaying effective treatments and potentially leading to more severe health outcomes. The ethical ramifications of such non-disclosure are profound. Withholding information that could prevent harm contradicts the foundational moral principle of non-maleficence, the commitment to do no harm, which should guide business practices just as it does medical practice.

The legal landscape offers some support for consumer rights but is a patchwork of regulations that can vary significantly by region. In the United States, for instance, the Federal Food, Drug, and Cosmetic Act mandates certain disclosures on food labels, such as ingredients and allergens, to protect consumers. However, these regulations often allow for vague descriptions like "natural" flavors" and "pieces," which" tell you little about the contents. Proposed laws that enhance transparency include stricter GMO labeling requirements, which have sparked intense debate about consumer rights versus corporate rights to intellectual property protection. These legislative efforts reflect a growing consumer demand for greater transparency in the food industry, advocating for the right to make informed choices about what they eat.

The ethical argument for prioritizing transparency over secrecy in the food industry hinges on the principle of beneficence — promoting the wellbeing of others. Manufacturers enable healthier choices and contribute to a greater societal good by providing precise and comprehensive information about food products. This transparency fosters trust and strengthens the relationship between consumers and producers. It reassures you that your food is safe, ethically sourced, and produced without harming others or the environment. Moreover, in an era where consumer loyalty increasingly depends on corporate social responsibility, transparency can enhance competition and competitive advantage, turning ethical practice into a profitable strategy.

In navigating these complex issues, it becomes apparent that the right to know often outweighs the need for secrecy. While protecting trade secrets is undoubtedly essential for fostering innovation and economic growth, this should not come at the cost of consumer safety and wellbeing. Ethical business practices, supported by robust legal frameworks, provide the foundation for a food industry that respects consumer rights while fostering a vibrant marketplace of ideas and products. As debates and policies evolve, the hope is for a fair and transparent food system that upholds the spirit and letter of ethical responsibility, ensuring that consumer trust is never compromised for corporate secrecy.

10.3 The Ethics of Bioengineering

Bioengineering in food production encompasses a broad spectrum of technologies, including genetic modification and synthetic biology. These innovations involve altering the genetic makeup of organisms to achieve desired traits such as disease resistance, nutritional enhancement, or increased yield. Genetic modification, for instance, can involve transferring specific genes from one organism to another to confer beneficial characteristics.

Ethical Considerations

In contrast, synthetic biology may involve designing organisms from scratch to perform particular functions.

The ethical debates surrounding these technologies are multifaceted and deeply polarizing. One fundamental concern is playing 'God'—altering natural organisms in ways that could have unforeseen consequences. Critics argue that tampering with nature at such a fundamental level carries inherent risks and challenges the traditional boundaries of human intervention in nature. Moreover, there are concerns about the long-term ecological impacts of releasing genetically modified organisms (GMOs) into the environment. Potential threats include:

- The disruption of local ecosystems.
- Unintended harm to non-target species.
- The reduction of biodiversity.

For example, a genetically modified crop resistant to pests might also harm beneficial insects or lead to the emergence of superweeds that could further necessitate chemical herbicides.

The ethical dimension of consumer autonomy and choice in the context of bioengineered foods is also significant. Consumers have a fundamental right to know what is in their food and how it is produced, which supports their ability to make informed choices consistent with their values and health needs. Clear labeling of GMOs is at the core of this issue. Without transparent labeling, consumers may unknowingly consume bioengineered foods, undermining their autonomy and leading to a breach of trust in the food system. This is particularly relevant for those with ethical or health-related concerns about GMOs. The debate over GMO labeling indicates broader worries about transparency and the public's need to be fully informed about the products they consume.

Internationally, the acceptance of and regulations governing bioengineered foods vary widely, reflecting diverse cultural attitudes and ethical considerations. In the European Union, for instance, significant public skepticism about GMOs has led to stringent regulations and labeling requirements. In contrast, GMOs are much more widely accepted in the United States. They are not subject to mandatory labeling under federal law, reflecting a different balance of ethical priorities and economic interests. These variations highlight that there are no universally agreed-upon ethical principles in food production; instead, they are shaped by cultural values, financial considerations, and public attitudes.

This global disparity raises essential questions about food production ethics and the cultural context shaping these norms. It challenges us to consider whether global standards for bioengineering practices exist or whether ethical considerations in food production should always be localized. Furthermore, it invites reflection on balancing scientific innovation in food production concerning diverse ethical viewpoints and ecological concerns.

As we navigate these complex ethical landscapes, it becomes evident that the decisions made in laboratories and boardrooms extend far beyond the confines of scientific and economic considerations—they touch upon fundamental moral questions about our relationship with nature, our responsibility towards future generations, and the rights of consumers to make informed choices. In this light, bioengineering is not just a technical challenge but a profound ethical endeavor that requires us to continually evaluate and adapt our practices in the face of evolving scientific knowledge and societal values.

Reflecting on the Ethics of Bioengineering

In conclusion, this exploration of the ethics of bioengineering in food production highlights the delicate balance between harnessing scientific advancements for human benefit and addressing the ethical, ecological, and social implications of these technologies. The discussions of genetic modification and synthetic biology reflect concerns about safety and biodiversity and broader philosophical debates about human intervention in nature and the limits of scientific endeavor.

As we move forward, the challenge lies in fostering an informed dialogue that embraces both the potential benefits of bioengineering and the ethical complexities it entails. This requires robust regulatory frameworks, transparent communication, and ongoing ethical reflection that engages all stakeholders—scientists, policymakers, and consumers. By doing so, we can ensure that advancements in food production feed the growing global population and respect our ethical obligations to current and future generations.

As we turn the page to the next chapter, we will delve deeper into the role of activism and advocacy in shaping food safety standards, exploring how individuals and organizations can influence policy and public awareness to promote a safer, more ethical food system.

NOTES

CHAPTER 11
ACTIVISM AND ADVOCACY

Imagine a world where the food on your table is guaranteed safe by regulatory bodies and a vigilant, informed, and active community. This chapter is a call to arms, urging you, the reader, to convert your concern about food safety into concrete action. The power of grassroots movements in shaping policy and public opinion cannot be overstated. From the civil rights movement to environmental campaigns, history is replete with examples of how collective citizen action has forged pathways to substantial societal change. In this chapter, we delve into how you can initiate and nurture a grassroots movement focused on enhancing food safety, ensuring that the meals on our tables do not come at the cost of our health.

11.1 Starting a Grassroots Movement for Food Safety

Identifying Core Issues and Goals

The first step in sparking a grassroots movement is identifying food safety issues that resonate most deeply with your community. Is the use of pesticides, GMOs, or transparency about food additives most concern your neighbors? Pinpointing these issues requires conversations, surveys, and listening to the stories of those around you. Once identified, setting clear, achievable goals is crucial. These goals should not only address the immediate concerns but also aim to foster long-term improvements in food safety. For instance, short-term goals include hosting educational workshops on understanding food labels, while long-term goals could aim at lobbying for local legislation on mandatory GMO labeling.

Building a Community Base

The backbone of any successful grassroots movement is its community. Building this community starts with engaging diverse stakeholders—households, schools, local businesses, and healthcare providers—all vested in safe food. Recruiting volunteers can begin within your immediate circle but should expand through social media, community bulletin boards, and local events. However, the glue that holds these volunteers together is a shared commitment to the cause. A collaborative environment is essential, as it encourages exchanging ideas and resources, ensuring that all voices are heard and valued. Regular meetings, clear communication channels, and inclusive decision-making processes are the pillars upon which successful community movements are built.

Organizing Effective Campaigns

With a solid community base, the next step is to launch campaigns that effectively highlight and address the identified food safety issues. Effective campaigning combines strategic planning, resource management, and community engagement. Begin by setting clear objectives for each campaign, whether raising awareness through educational programs or influencing policy through petitions. Planning should include detailed timelines, assigned responsibilities, and budgeting. Public speaking events, food safety fairs, and school partnerships can serve as platforms to engage the wider community and amplify your message. Remember, every campaign should have a clear call to action, encouraging community members to participate actively.

Sustaining Momentum

Maintaining the momentum of a grassroots movement is often more challenging than starting one. It requires continuous effort, adaptability, and persistence. Keeping the community engaged with regular updates, success stories, and new challenges can help sustain interest and involvement. Financial sustainability can be achieved through fundraising events, grants, or sponsorships, essential for supporting ongoing activities. Moreover, adapting strategies based on feedback and the changing dynamics of the community ensures that the movement remains relevant and practical. Continuous education about new food safety research and regulatory changes can also keep the community informed and motivated.

Visual Element: Community Engagement Blueprint

An infographic titled "Community Engagement Blueprint" visual" summarizes effective strategies for building and sustaining a grassroots movement. It outlines critical steps in community building, campaign organization, and momentum sustenance, providing a quick and engaging reference for community activists.

As we continue to explore the multifaceted approaches to advocacy and activism in food safety, it becomes clear that the power to instigate change lies in policymakers and informed and active citizens like you. Armed with knowledge, driven by purpose, and united in action, grassroots movements can potentially transform public health landscapes. Let this chapter serve as your blueprint for action, guiding you through the practical steps of mobilizing your community toward a safer, healthier future.

11.2 Effective Lobbying Techniques for the Everyday Citizen

Navigating the complex terrain of legislative processes is essential for anyone looking to influence food safety policies effectively. Legislation, whether at the local, state, or federal level, follows a specific journey from conception to enactment. Initially, a bill is drafted, often by a lawmaker or a lawmaker, which addresses a particular issue or change in policy. This proposal is then introduced to the legislative body, which may be assigned to one or more committees specializing in relevant areas, such as agriculture or public health. These committees review and possibly amend the proposal before deciding whether to send it back to the floor for all members to consider. Suppose the bill passes through all legislative stages, including potential debates, votes, and possibly reconciliations between different legislative branches. In that case, it is then sent to the executive branch, where it can be signed into law or vetoed. Understanding this process helps you pinpoint where your advocacy can have the most significant impact, particularly during committee reviews where the details of legislation are scrutinized and shaped.

Crafting a compelling message is crucial in lobbying for change. This message should be clear, concise, and tailored to resonate with different audiences. When addressing legislators or regulators, you must base your arguments on solid data and articulate the tangible benefits of proposed changes, such as enhancing public health or reducing healthcare costs. For the general public, your message might focus more on personal stories and the direct impact of food safety issues on families and communities, which can be more engaging and relatable. Effective messaging informs and evokes a sense of urgency and need for action, encouraging stakeholders to advocate for the necessary changes alongside you.

Engaging with lawmakers is more than just sending emails or making phone calls; it involves building relationships and presenting your case in ways that align with their interests and constituents. Scheduling face-to-face meetings in local district offices or state/national capitols can be particularly effective. Prepare for these meetings by clearly outlining your objectives and familiarizing yourself with the lawmakelawmaker'soting record and public statements on food safety. During the meeting, be concise and focus on how the proposed changes will benefit your interests and those of the broader community. Follow-up is equally crucial; sending thank-you notes reiterating key points can keep the conversation going and maintain the relationship for future advocacy.

Building coalitions with other advocacy groups can significantly amplify your lobbying efforts. Coalitions unite diverse groups with common interests, pooling resources, knowledge, and influence to create a stronger voice for change. These partnerships can vary from informal alliances that share information and coordinate activities to more formal arrangements with joint funding and dedicated staff. Working in a coalition allows for sharing the load of lobbying work, from tracking legislation to speaking at public hearings, and can extend your reach to more stakeholders and decision-makers. Moreover, a united front of multiple organizations advocating for the same changes can be more persuasive, demonstrating broad-based support that can sway policymakers.

Remember that persistence and adaptability are your allies in enhancing food safety through effective lobbying. Legislative change often requires sustained effort over multiple sessions and adapting strategies in response to political and social shifts. Your advocacy can lead to safer food practices that benefit everyone, reinforcing the importance of citizen involvement in legislative processes and the power of informed coordinated action.

11.3 Using Social Media for Advocacy and Change

In today's age, social media platforms are tools for social interaction and powerful instruments for advocacy and change, especially in realms as critical as food safety. Selecting the right one based on your advocacy goals is crucial to leverage these platforms effectively. Consider the demographics of your target audience: platforms like Facebook have a broad, somewhat older user base, making it ideal for widespread outreach, whereas Instagram and TikTok, with their younger audiences, are perfect for engaging through visual content and quick, impactful messages.

Additionally, Twitter can be highly effective for real-time updates and engaging directly with stakeholders and policymakers. Each platform has unique features that can be strategically used, such as Instagram Stories for daily updates or Facebook Groups to foster community discussions.

Creating content that resonates with your audience is vital. It's more than just sharing facts; it's about telling stories that connect emotionally. Use compelling visuals like infographics that distill complex information into digestible, shareable content. Videos can be particularly impactful, whether detailed explanations of food safety issues or heart-touching stories from individuals affected by foodborne illnesses. Every piece of content should have a clear, engaging, and relatable message that underscores the importance of your advocacy, encouraging viewers to share, comment, and become active participants in the cause. Remember, the goal is to inform, engage, and inspire action, not just to broadcast information.

Building an online community fosters a space where interactions are meaningful, and everyone feels valued. This starts with regular engagement — responding to comments, participating in discussions, and showing appreciation for support. Moderation is vital; establish policies that encourage respectful and constructive dialogue and quickly address any negative behaviors that could disrupt the community. Organize regular online events or live sessions, which can help strengthen community ties, keeping the members engaged and motivated. Featuring stories from community members, sharing behind-the-scenes content, or hosting Q&A sessions can make your social media space more inclusive and dynamic.

To gauge the effectiveness of your social media advocacy, utilize analytics tools provided by the platforms. These tools can offer insights into who is engaging with your content, which types of content perform best, and what times your audience is most active. This data is invaluable for refining your strategies. For example, if videos on specific topics gain more interaction, consider increasing video content in your planning. Regularly reviewing this data allows you to adapt your approach effectively, ensuring your efforts are as impactful as possible. Adjustments based on analytics can significantly increase your reach and engagement, amplifying your advocacy for food safety.

As we close this chapter on utilizing social media for advocacy and change, remember that these platforms are gateways to vast networks of people who can join your cause. By choosing the right platforms, creating resonant content, fostering a supportive community, and continuously optimizing your strategies based on analytics, you can transform your digital presence into a dynamic force for advocacy in food safety. As we move into the next chapter, we'll examine how these digital engagements translate into real-world actions and the broader implications of these movements in shaping policies and public perceptions.

CHAPTER 12
THE FUTURE OF FOOD

In the ever-evolving food safety landscape, harnessing advanced technologies promises a revolution in safeguarding our food supply. Imagine a world where the outbreak of foodborne illnesses is predicted before it even happens, an ever-watchful digital eye monitors compliance with food safety norms, and every morsel of Food on your plate can be traced back to its source with a few clicks. This is not science fiction; it is the future of food safety, propelled by integrating Artificial Intelligence (AI) into food monitoring systems. As we venture into this new era, understanding AI's capabilities, implications, and ethical considerations in ensuring our Food remains safe is paramount.

12.1 The Role of AI in Food Safety Monitoring

Automating Risk Assessment

Artificial Intelligence, with its unparalleled ability to analyze vast datasets quickly and precisely, redefines risk assessment in food production. Traditional methods, often cumbersome and time-consuming, can only react to outbreaks. In contrast, AI-driven systems proactively predict and prevent potential health hazards. By integrating data from multiple sources—weather patterns, production processes, historical outbreak information—AI algorithms can identify risk factors that precede contamination events. For instance, AI can detect subtle changes in storage temperatures or humidity levels that might promote bacterial growth, prompting preemptive action. This shift from reactive to preventive measures could dramatically reduce the incidence of foodborne illnesses, saving lives and resources.

AI in Inspection and Compliance

The role of AI extends beyond risk assessment to enforcing food safety standards through continuous monitoring. Equipped with AI, cameras, and sensors, food production facilities can analyze real-time video feeds and instantly identify deviations from required practices. This might include detecting if workers fail to sanitize their hands or if a product is processed at incorrect temperatures. These AI systems provide an ongoing audit of compliance that is far more rigorous and comprehensive than periodic human inspections. The result is a consistently upheld food safety standard, ensuring that the Food reaching your table has been handled with the utmost care.

Enhanced Traceability and Response

One of the critical challenges in managing food safety incidents is tracing the contamination source quickly. AI significantly enhances traceability in the food supply chain. By analyzing data points across the chain—from farm to fork—AI can soon pinpoint

The Future of Food

the contamination source, be it a specific batch of crops or a particular production line. This rapid identification facilitates swift action to remove the contaminated product from the market and minimizes the disruption to the food supply chain, ensuring that safe, healthy Food remains available. This enhanced traceability is not just about managing crises; it also fosters consumer confidence as people become more aware of the robust mechanisms to protect their Food.

Challenges and Ethical Considerations

However, integrating AI into food safety systems is full of challenges. Privacy concerns emerge as extensive data collection becomes a norm. The transparency of AI algorithms—how decisions are made and on what basis—is another significant concern. There is a risk that reliance on AI could lead to a 'black box' scenario, where decisions are not easily interpretable by humans. Moreover, the displacement of human workers by AI systems poses ethical and economic challenges. Balancing these concerns with the benefits of AI requires robust regulatory frameworks, ongoing oversight, and a commitment to ethical AI practices. Ensuring that AI systems are transparent and their decision-making processes are understandable is crucial to maintaining trust in these advanced technologies.

Visual Element: AI in Food Safety Flowchart

An illustrative flowchart detailing how AI integrates into various stages of food production and monitoring is provided to better understand AI's role in food safety. It visually represents the process from data collection, risk analysis, real-time monitoring, enhanced traceability, and rapid response, offering a clear view of how AI creates a tightly controlled safety net across the food supply chain.

As we look to the future, the potential of AI to transform food safety is immense. With careful implementation, consideration of ethical issues, and rigorous standards, AI could make our Food safer and improve the efficiency and transparency of the food supply chain. As you continue navigating through the aisles of your local grocery store, take comfort in knowing that the future of food safety is promising, where technology stands as a vigilant guardian of public health.

12.2 The Potential of Blockchain in Tracking Food Ingredients

Blockchain technology, often associated with cryptocurrencies, is a secure and transparent way to record transactions on multiple computers. The record cannot be altered retroactively without altering all subsequent blocks and the network consensus. This technology offers a high level of security as it makes it extremely difficult to tamper with data. The decentralized nature of blockchain provides this robust security and transparency, making it an excellent solution for tracking the food industry's intricate and often opaque supply chains.

When applied to food safety, blockchain can revolutionize how we track and verify every step of the food supply chain—from the farmer's field to the grocery store shelf. By creating a tamper-proof record of each transaction, blockchain technology can provide a comprehensive, immutable ledger that all parties in the supply chain can access. Every time a food item is passed along the supply chain, a transaction record is logged in the blockchain, including details like time, date, location, quality, and safety data. This level of traceability enhances transparency and significantly improves food safety by making it easier to track the origin of contamination or verify the product's authenticity. For instance, if a batch of lettuce is found to be contaminated, retailers can quickly trace back through the blockchain ledger to identify all supply chain movements of that batch right back to the farm of origin, potentially within minutes. This rapid traceability can dramatically reduce the health impact of foodborne disease outbreaks and the economic cost to food suppliers and retailers.

Cases of Blockchain in Action

Several pioneering cases have demonstrated the successful implementation of blockchain in the food industry. A notable example is a leading global retailer collaborating with a technology giant to apply blockchain to its food supply chains. The project initially focused on tracing pork in China and produce in the US. The system allowed the company to trace each piece of Food from the farm through every stage of processing, handling, and selling, ensuring that any food safety issues could be addressed and managed quickly and precisely. Another instance is a famous coffee company that uses blockchain to trace the journey of its coffee beans from farms in Costa Rica, Colombia, and Rwanda to its cafes. Customers can scan codes on the coffee bags to access this information, enhancing consumer trust and engagement by providing proof of authenticity and ethical sourcing.

Scaling and Adoption Challenges

Despite its potential, the widespread adoption of blockchain in the food industry faces several challenges. One of the primary barriers is the technological complexity and the significant shift in operations that blockchain implementation requires. Many food companies are accustomed to traditional record-keeping methods and may need help transitioning to a high-tech solution. Additionally, for blockchain to be truly effective, it needs to be adopted across the entire supply chain, which requires coordination and collaboration among numerous stakeholders, including farmers, suppliers, processors, distributors, and retailers, who may have varying levels of technological literacy and resources.

Another significant challenge is the establishment of industry-wide standards for blockchain applications in food safety. Standardized protocols and technologies can make it easier for different systems to interact with each other, which is essential for the seamless tracking of products across diverse supply chains. Moreover, some stakeholders resist blockchain's transparency, potentially exposing sensitive business information to competitors.

As we continue to explore the transformative potential of blockchain in ensuring the safety and integrity of our Food, it is clear that this technology holds the promise of building a more transparent, secure, and efficient food supply chain. However, realizing this potential will require overcoming significant technical and collaborative hurdles. The food industry must be prepared to embrace these challenges, pushing forward with innovation and cooperation to make the benefits of blockchain a standard practice in food safety management.

12.3 Innovations in Non-Toxic Food Preservation

The quest for safer and more sustainable food preservation methods is gaining momentum, driven by increasing consumer demand for products free from synthetic chemicals and the global push towards environmental sustainability. Innovations in non-toxic food preservation, such as advanced packaging solutions and natural plant-derived preservatives, are at the forefront of this revolution, offering promising alternatives to traditional methods that rely heavily on chemical additives. These emerging techniques extend the shelf life of Food and maintain its quality and nutritional value, presenting a win-win situation for both consumers and producers.

Advanced packaging solutions, for example, employ biodegradable materials and active packaging technologies that interact with Food to slow down spoilage processes. These might include films infused with natural antimicrobial agents like essential oils or packets that absorb excess moisture or oxygen, the primary culprits in food degradation. Such innovations reduce reliance on chemical preservatives and align with the growing environmental consciousness among consumers by offering more sustainable disposal options. Meanwhile, the resurgence of interest in natural preservatives—extracts from herbs, spices, and fruits known for their antimicrobial properties—reflects a shift towards embracing nature's solutions for food preservation. These substances, often rich in beneficial antioxidants, offer dual advantages by extending shelf life and potentially enhancing the nutritional profile of Food.

The benefits of transitioning to non-toxic methods of food preservation extend beyond just keeping Food fresh. These methods inherently promote better health outcomes by reducing exposure to harmful chemicals. Consumers are becoming more aware of the links between synthetic additives and various health issues, which drives the demand for cleaner labels. Environmentally, the shift towards biodegradable packaging and naturally derived preservatives significantly reduces the ecological footprint of food production. Less reliance on synthetic chemicals means fewer environmental pollutants, and biodegradable packaging materials help minimize waste and resource consumption.

Research and development in this field are vibrant, reflecting a collaboration across various sectors. Academic institutions, startup companies, and established food corporations are increasingly investing in research to discover and refine new methods of non-toxic preservation. These efforts are supported by technological advancements that enable better isolation and characterization of natural compounds with preservative properties. Furthermore, the application of nanotechnology and biotechnology is expanding the possibilities for enhancing the efficacy and stability of natural preservatives, paving the way for their broader commercial use.

However, the widespread adoption of these innovative preservation methods is contingent upon consumer acceptance and market trends. Tastes, cost, and perceived safety are critical in consumer decisions. Natural preservatives and advanced packaging solutions must meet or exceed traditional methods' sensory and convenience standards to gain significant market traction. Additionally, transparent communication regarding the benefits and safety of these new methods is essential to dispel any misconceptions and build consumer trust. As research continues to affirm the safety and efficacy of these innovations, consumer confidence will grow, leading to greater acceptance and preference for products utilizing non-toxic preservation methods.

In the grand scheme of food safety and sustainability, the innovations in non-toxic food preservation mark a significant step forward. They offer safer, healthier food options and contribute to the broader environmental goals of reducing chemical use and waste. As this chapter concludes, we see a clear trajectory toward a more sustainable and health-conscious food system driven by technological innovation and consumer empowerment. These developments are poised to reshape how we preserve Food and consider our food systems' safety and sustainability.

Innovations in non-toxic food preservation are setting the stage for a revolution in approaching food safety and environmental sustainability. The food industry is moving towards safer, healthier, and more sustainable practices by embracing advanced packaging solutions and natural preservatives. These developments respond to consumer demands for cleaner labels and less environmental impact, and they offer improved health benefits by reducing exposure to harmful chemicals. As research progresses and market acceptance grows, these innovative methods are expected to become more prevalent, leading to broader changes in food production practices. Moving forward into the next chapter, we will explore the role of consumer education and regulatory frameworks in supporting these advancements, ensuring they realize their full potential in transforming our food systems.

NOTES

CHAPTER 13
BUILDING RESILIENCE

To navigate the food industry's complex and often murky waters, we must arm ourselves with knowledge and the tools to maintain our psychological and emotional wellbeing. As we peel back the layers of what truly goes into our Food, the revelations can sometimes be overwhelming, even distressing. This chapter is designed to anchor the storm, offering strategies to stay informed while safeguarding your mental health.

13.1 Staying Informed Without Becoming Overwhelmed

The modern media landscape bombards us with a constant stream of information, and while staying informed is vital, there is a fine line between being well-informed and becoming overwhelmed. This is particularly true in food safety, where the stakes are high, and the news can often be unsettling. To navigate this deluge of data without drowning in it, it's essential to set limits on your news consumption. Consider designating specific times of the day to catch up on the latest developments rather than allowing it to be a constant influx that can heighten anxiety and distract from daily responsibilities. Choose news sources that are known for their credibility and factual reporting. Organizations like the Center for Science in the Public Interest (CSPI) or the Environmental Working Group (EWG) provide research-based information on food safety and public health that can be trusted for accuracy and relevance.

Critical thinking is your most valuable tool in distinguishing between fact and fiction. In food safety, misinformation can spread as rapidly as fact, often sensationalized to draw clicks and views. Develop the habit of questioning what you read and hear: Who is the source? What is their motive? Are they citing scientific evidence, and if so, how robust is it? You can prevent misinformation from clouding your understanding and decision-making processes by approaching information with a discerning eye.

Building emotional resilience is equally essential, mainly as you uncover the less palatable realities of the food industry. Techniques such as cognitive reframing can be powerful here. This involves consciously shifting your perspective on the information you encounter. For instance, instead of feeling defeated by reports of food safety violations, view these revelations as empowering knowledge that enables you to make safer choices. Stress-reduction practices such as meditation, deep breathing exercises, and regular physical activity can also effectively manage the emotional toll that sometimes accompanies deep engagement with challenging subjects.

Engaging with community groups or online forums focused on food safety can provide a sense of solidarity and shared purpose that mitigates feelings of helplessness or isolation. These platforms offer spaces to exchange information, share personal experiences, and collaborate on initiatives that promote better food safety practices. Whether through local community meetings or online groups, connecting with others who share your concerns and commitment can reinforce your efforts and provide the emotional support necessary to continue advocating for change.

Visual Element: Infographic on Balancing Information Intake

An infographic illustrates strategies for balancing staying informed with maintaining mental health to further aid in managing your information intake. This visual guide includes tips on scheduling news consumption, vetting sources for credibility, and employing critical thinking. It serves as a quick reference tool that you can turn to whenever the flow of information feels overwhelming, helping you stay grounded and focused.

By equipping yourself with these strategies, you can remain an informed and active participant in the dialogue and advocacy surrounding food safety without letting the weight of this knowledge detract from your quality of life or mental health. As we continue to explore the complexities of the food industry, remember that resilience is not just about enduring but also about thriving amidst challenges. With the right tools and support, you can navigate this journey confidently and clearly, contributing to a safer, more transparent food system for all.

13.2 Supporting Local and Organic Food Sources

In an age where the origins of our Food are often a mystery, turning to local and organic sources offers a beacon of transparency and trust. By choosing local and organic foods, you're making a decision that benefits your health and contributes to environmental sustainability and the local economy. These food sources generally use fewer pesticides, reduce transportation emissions due to shorter distribution chains, and support farming practices that maintain soil integrity and biodiversity. Moreover, the economic benefits are palpable—local purchases help keep money within the community, supporting local farmers and creating jobs.

Finding and accessing local and organic foods is simpler than you think. Farmers' markets are treasure troves of fresh, seasonal produce directly from the growers. Here, you can find not only fruits and vegetables but often meats, cheeses, and other artisan products that are organic and sustainably produced. Another fantastic way to immerse yourself in the local food scene is by joining a Community Supported Agriculture (CSA) program. In a CSA, you buy a "share" of produce from a local farm, and every week or bi-weekly, you receive a box of fresh produce. This ensures you get the freshest seasonable vegetables and fruits and connects you directly with farming. For those with a bit of land or even a sunny spot on a balcony, starting a home garden can be a rewarding way to ensure you have access to organic Food. Simple and immensely gratifying, gardening can be a profound step towards self-sufficiency and understanding the food process from seed to plate.

Building relationships with local farmers and producers can enrich your understanding of where your Food comes from and the effort that goes into producing it. These relationships are foundational to appreciating the challenges and intricacies of sustainable agriculture. Many farmers are more than willing to share their knowledge about their farming practices, providing insights into their challenges, from weather conditions to market prices. These interactions can transform how you think about Food and your choices at the grocery store. They humanize the food process, reminding us that behind every vegetable or piece of fruit is a network of people and a wealth of hard work and dedication.

Advocating for greater availability and support of local and organic foods in your community can start with simple actions like requesting local products at your grocery store or school cafeteria. Organizing or joining a community food co-op can amplify this advocacy, creating an influential collective that directly supports local farmers and organic practices. Co-ops provide members with fresh, regional, and organic produce, strengthen community bonds, and promote a culture of sustainability and health. By championing these initiatives, you play a crucial role in transforming local food systems and fostering an environment where sustainable and healthy food choices are accessible to everyone in your community.

Navigating the shift to local and organic foods is not just about personal health—it's about participating in a broader movement toward sustainability and ethical consumption. This food-sourcing approach allows you to control your dietary choices while supporting practices in harmony with the environment and local economies. As you continue to engage with local farmers, explore CSAs, and perhaps even start your garden, remember that every small choice adds up to a significant impact, paving the way for a healthier, more sustainable world.

13.3 Mindfulness and Dietary Choices

Mindfulness, a practice deeply rooted in being present and fully engaged, has profound implications when applied to eating. It's about experiencing Food more intensely and making conscious choices that enhance health and wellbeing. When you eat mindfully, you slow down, savor each bite, and listen to your body's hunger and fullness signals. This practice increases the enjoyment of Food and encourages healthier eating habits by reducing mindless overeating and impulsive food choices. For instance, by focusing entirely on the experience of eating, you're more likely to notice when you are satisfied, helping to prevent overeating. Additionally, mindfulness helps recognize the triggers for unhealthy eating behaviors, such as stress or boredom, and provides strategies for managing them more effectively.

The impact of mindfulness extends beyond personal health to influence broader dietary choices. When you're mindful, you tend to make food choices that are not only good for you but also ethically sound and environmentally sustainable. This might mean opting for organically grown, locally sourced, or not excessively packaged foods. Such decisions often stem from a heightened awareness of how food choices impact the environment, animal welfare, and labor practices. By being mindful, you start to see the connection between your immediate food choices and their more significant effects, guiding you towards decisions that align with your values about health, sustainability, and ethical responsibility.

In addressing the complexities of the food industry, mindfulness offers a powerful tool for navigating information and making informed choices without becoming overwhelmed. The practice encourages a calm, reflective approach to consuming information about food sources, processing, and industry practices. This can help maintain a balanced perspective, avoid sensationalism's pitfalls, and focus on actionable knowledge. For example, when faced with conflicting reports about food safety or nutritional guidelines, a mindful approach helps you to evaluate the information critically, reflect on its sources, and decide how it fits into your more extensive understanding of Food and health.

Various resources and techniques are available to cultivate mindfulness in your eating habits. Guided meditations can train your attention and awareness, helping you bring these skills to the dinner table. Journaling about your eating experiences can deepen your understanding of your dietary habits and the emotions or situations that influence them. Educational workshops, often available through community health centers or online, can provide practical tips and support for integrating mindfulness into your daily routine. These resources enrich your practice and connect you with a community of like-minded individuals who share your commitment to mindful eating.

Incorporating mindfulness into your dietary choices transforms eating from a routine task into a deliberate practice that nurtures your body, mind, and the environment. As you become more attuned to the implications of your food choices, you contribute to a demand for more transparent, ethical, and sustainable food systems. This chapter has explored mindfulness's decisive role in reshaping how we think about and engage with our Food. By adopting mindful eating practices, you can enjoy more affluent, more satisfying eating experiences while making choices that promote a healthier, more equitable world.

As we conclude this exploration into mindfulness and dietary choices, remember that each meal presents an opportunity to practice mindfulness. This approach doesn't just change how you eat; it transforms your relationship with Food, leading to enhanced health, a deeper connection to your environment, and a positive impact on the global food system. The next chapter will explore how these mindful practices can be extended beyond personal habits to influence community and policy levels, further amplifying the benefits of mindful eating in creating sustainable and ethical food landscapes.

NOTES

CHAPTER 14
GLOBAL MOVEMENTS

In a world increasingly dominated by fast-paced lifestyles and the ubiquity of fast food chains, a counter-movement born in the rolling hills of Italy's Piedmont region stands out as a beacon of resistance and hope. In the late 1980s, it was here that the Slow Food movement first took root, sprouting from the passionate protests against opening a McDonald's near the iconic Spanish Steps in Rome. This movement, founded by Carlo Petrini and a group of activists, was fueled by the desire to preserve local food cultures and traditions, which were rapidly eroded by the homogenizing forces of fast Food and global capitalism. The core philosophy of Slow Food is beautifully simple yet revolutionary: to counter the rise of fast life and combat people's dwindling interest in the Food they eat, where it comes from, how it tastes, and how our food choices affect the world around us.

14.1 The Slow Food Movement: A Global Perspective

Origins and Philosophy of the Slow Food Movement

The inception of the Slow Food movement was marked by an ardent commitment to the pleasures of the table combined with a dedication to sustainability and biodiversity. It advocates for a food system that is "good, clean, and fair," promoting delicious Food that is produced in harmony with the environment and providing fair wages to those who make it. This philosophy not only aims to preserve endangered species of domestic animals and wild Food varieties but also strives to maintain traditional methods of cultivation and food production that are at risk of extinction due to the industrialization of agriculture and the consolidation of food production into the hands of a few multinational corporations.

Impact on Local and Global Food Policies

Slow Food's influence has percolated through local and global food policies over the decades. At the regional level, Slow Food chapters work closely with small-scale producers and artisans to protect local ecosystems and food traditions, enhancing local economies and promoting sustainable agriculture. Internationally, Slow Food has been a vocal advocate in global forums against genetically modified organisms (GMOs) and the monopolization of seeds by large agrochemical companies. Their efforts have led to greater awareness and stricter regulations on food production and sustainability standards across Europe and beyond. Slow Food has reshaped how food safety and quality are perceived and legislated globally by lobbying for policies prioritizing biodiversity and ecological sustainability.

Expansion and Global Reach

From its humble beginnings in Italy, Slow Food has burgeoned into a global movement, boasting over a million supporters in more than 160 countries. Each national chapter tailors Slow Food's goals to local contexts, addressing specific food sovereignty, sustainability, and heritage issues. This global network helps promote the movement's core messages and manipulates its strategies to fit diverse cultural perspectives on Food and agriculture. The movement's biennial Terra Madre event in Turin, Italy, exemplifies this, gathering food communities worldwide to share knowledge, celebrate food diversity, and forge alliances.

Case Studies of Successful Initiatives

One of the most impactful initiatives of Slow Food has been the Ark of Taste project, an international catalog that identifies and catalogs endangered heritage foods from across the globe. This project raises awareness about the richness and diversity of culinary traditions and helps protect them by encouraging farmers and consumers to grow and seek out these rare products. Another significant achievement is the Terra Madre network, which links small-scale food producers worldwide with chefs, academics, and activists to foster a sustainable food system. These examples underscore how local action can influence global trends, promoting a food system that respects the environment, human dignity, and health.

Visual Element: Map of Slow Food's Global Impact

A detailed map highlights the countries with active Slow Food chapters and critical projects implemented in these regions to visually depict the movement's expansive reach and influence. This map illustrates the movement's widespread popularity and showcases the variety of its initiatives, from preserving ancient grains in Morocco to promoting sustainable seafood in Japan.

As you reflect on the transformative power of the Slow Food movement, it becomes evident that change, although gradual, is profoundly possible when communities unite with a common purpose. This global network continues to inspire and effectuate change, advocating for a world where everyone has access to Food that is good for them, those who grow it, and the planet.

14.2 International Coalitions for Food Safety

In a world where food production and consumption cross multiple borders, the necessity for international collaboration in food safety is undeniable. Prominent coalitions such as the Codex Alimentarius Commission and the Global Food Safety Initiative (GFSI) have risen to the forefront of this global effort, orchestrating harmonious standards and practices across nations. The Codex Alimentarius Commission, established in 1963 by the Food and Agriculture Organization (FAO) and the World Health Organization (WHO), is a principal entity in this domain. Its main objective is to develop harmonized international food standards that protect consumer health and ensure fair practices in food trade. Similarly, the Global Food Safety Initiative, launched in 2000, emerged from the consumer goods industry's realization of the need for enhanced food safety throughout the supply chain.

Its mission pivots to benchmarking food safety standards, improving cost efficiency throughout the food supply chain through common standards, and fostering a mutual acceptance of safe food handling practices.

These coalitions' role in harmonizing international food safety standards cannot be overstated. By providing a platform where countries can collaborate and negotiate, they help standardize regulations and guidelines that ensure the safety and quality of food products globally. This harmonization is crucial not only for consumer safety but also for facilitating international trade. A product manufactured in one country and consumed in another can adhere to universally recognized safety standards, simplifying global commerce and reducing barriers to entry in different markets. For instance, the Codex Alimentarius operates a series of standards, guidelines, and codes of practice that are recognized as reference points for food safety under the World Trade Organization's Sanitary and Phytosanitary Agreement. This universal benchmark reduces the complexity and cost of compliance for food exporters while maintaining high safety standards for consumers worldwide.

Highlighting specific achievements, these coalitions have been instrumental in several global food safety improvements. The Codex Alimentarius Commission, for example, has developed guidelines for controlling Campylobacter and Salmonella in chicken meat, two of the most common bacterial causes of foodborne diseases globally. These guidelines have been pivotal in helping countries control these pathogens, thereby protecting the health of millions. The Global Food Safety Initiative has significantly advanced the implementation of food safety management systems across the food supply chain. Through its benchmarking model, it has elevated numerous local food safety standards to meet international expectations, facilitating safer food practices worldwide.

However, the path forward is full of challenges. One of the primary challenges these coalitions face is the need for more implementation of food safety standards across countries, particularly between developed and developing nations. While developed countries often have the resources and infrastructure to implement rigorous food safety standards, developing countries may need more resources, infrastructure, and technical expertise. This disparity can lead to uneven food safety levels worldwide and pose significant challenges in international food trade. Moreover, there is a pressing need for greater inclusion of developing nations in these international coalitions' development and decision-making processes. Ensuring that these countries are not just participants but active contributors can enhance the relevance and applicability of the standards being developed.

To enhance global food safety collaboration effectively, future directions may include increasing technical and financial support for developing nations to build their food safety capacities. Leveraging technology and innovation can also play a crucial role in bridging gaps in food safety practices globally. For instance, deploying blockchain technology for traceability can improve transparency and accountability in the food supply chain, making tracking and managing food safety risks easier. Additionally, fostering a more inclusive approach in the governance of these coalitions can ensure that the standards developed are universally applicable and sensitive to the diverse economic, cultural, and agricultural landscapes across the globe.

As these coalitions continue to forge paths towards safer global food systems, their role in shaping an integrated, universally safe, and equitable food supply chain becomes increasingly critical. By addressing existing challenges and focusing on inclusive, technology-driven strategies, these international coalitions stand poised to significantly enhance food safety standards worldwide, ensuring that they adhere to the highest safety standards no matter where Food is produced or consumed.

14.3 Case Study: The Indian Organic Food Revolution

The surge in organic farming in India is not just a trend but a robust movement toward sustainable and health-conscious agricultural practices. Tracing back to the early 2000s, the organic movement gained momentum as Indian farmers began to rediscover traditional farming methods that forego synthetic chemicals. This shift was driven by environmental, health, and economic factors. Increasingly aware of the detrimental impacts of chemical fertilizers and pesticides on land and health, farmers and consumers pushed for a return to practices that ensure long-term soil sustainability and more significant health benefits. The growth of this sector has been remarkable, with India now ranked as one of the top countries in terms of area under organic cultivation and number of producers. This expansion into organic farming has transformed large swathes of India's agriculture into more sustainable practices and made significant inroads into international markets, catering to the global demand for organic products.

The Indian government and various non-governmental organizations (NGOs) have played a pivotal role in this transformation. Recognizing the potential of organic farming to support sustainable development, the government has implemented several initiatives to promote organic agriculture. Programs like the Paramparagat Krishi Vikas Yojana (PKVY) aim to assist farmers in adopting organic cultivation practices by providing them with financial assistance and training. These efforts are supplemented by NGOs that work at the grassroots level to educate farmers about the benefits of organic farming and provide technical support. Furthermore, the government's move to establish the National Programme for Organic Production (NPOP) has set standards for organic products and streamlined the certification process, enhancing consumer trust and export opportunities.

The impact of this shift towards organic farming on Indian farmers and local communities has been profound. Economically, it has opened new avenues for farmers by tapping into the lucrative market of organic goods, both domestically and internationally. This has benefited smallholder farmers, who find better and more stable pricing in organic markets than traditional farming. Health-wise, communities have noted decreased health issues related to pesticide exposure daily in conventional agricultural practices. Environmentally, organic agriculture has contributed to better soil health, reduced water pollution, and increased biodiversity in farming regions, which are critical in climate change and soil degradation.

However, the transition to organic farming has been challenging. One of the major hurdles is the high cost of organic certification, which can be prohibitive for small-scale farmers. Although government subsidies help, the process remains cumbersome and expensive. Access to markets is another significant challenge, as organic products often require separate supply chains that are better developed than conventional ones. Additionally, competition with traditional farming practices, which yield higher outputs due to chemical enhancers, poses an ongoing challenge for organic farmers who must achieve comparable yields using natural methods.

Despite these challenges, the organic farming revolution in India provides valuable lessons for other countries looking to transition to sustainable agricultural practices. The importance of government and community support is evident, as is the need for robust systems that facilitate market access and make certification more accessible to small farmers. Moreover, it underscores the necessity of educating farmers and consumers about the benefits of organic farming, which is crucial for this transition.

NOTES

CHAPTER 15
CALL TO ACTION

As we wrap up this exploration of the Indian organic food revolution, we see a vibrant tapestry of tradition blending with modern sustainability efforts. This movement enhances the health of the land and its people. It sets a precedent for global agricultural practices, highlighting the profound impact of concerted local and national efforts in fostering a sustainable and healthful food system. The insights from India's experience could illuminate paths for other nations striving to cultivate more sustainable and equitable food landscapes.

In the labyrinth of our daily grocery shopping, hidden behind the bright packaging and enticing marketing slogans, lies a complex web of food labels that often conceal more than they reveal. In the fine print, it's here that the battle for transparency in our food system continues. As someone deeply concerned about the integrity of what we consume, understanding how to navigate this maze is not just beneficial—it's imperative. This chapter is dedicated to empowering you with the knowledge to decipher food labels accurately, ensuring that your choices align with your values and health priorities.

15.1 How to Read and Understand Food Labels Accurately

Decoding Label Elements

Every element on a food label, from the nutritional facts to the list of ingredients, is critical to unlocking the secrets of what's in our Food. The nutritional facts panel provides a snapshot of the essential nutrients and their quantities in each serving. This includes calories, fats, carbohydrates, proteins, and vitamins, which are crucial for managing dietary needs and health conditions. However, the real art lies in interpreting these figures in the context of your daily nutritional requirements and understanding how they contribute to or are detailed from a healthy diet.

The ingredient list reveals the product's contents in descending order by weight. This means that the first listed ingredients are in the most significant amounts. Here, vigilance is required to spot undesirable additives or hidden sugars. Names like 'sucrose,' 'high-fructose corn syrup,' or 'inverted sugar' all point to added sugars, which can be detrimental in excess. Similarly, allergen warnings are crucial for those with specific food sensitivities or allergies, as they indicate the presence of common allergens like nuts, dairy, or gluten.

Understanding these elements isn't just about making healthier choices; it's about taking control of your diet and, by extension, your health and wellbeing. It's about not letting the food industry dictate your consumption through obfuscation and confusion.

Identifying Misleading Claims

Food labels are often adorned with claims that can be misleading. Phrases like 'all-natural,' 'low-fat,' or 'supports immunity' can create a health halo effect, suggesting that the product benefits health without substantial backing. For instance, a 'low-fat' product might still be high in sugars and overall calories, offsetting any benefits from reduced fat content. Learning to see through these marketing tactics is essential. Real-life examples abound; for instance, products labeled 'made with real fruit' may contain only trace amounts of fruit but high levels of artificial flavors and sugars. By understanding these tactics, you become better equipped to make good choices on paper and for your body.

Understanding Food Additives

Food additives are the most cryptic components of any ingredient list. Used for various purposes like preservation, coloring, and texture enhancement, these substances can sometimes be harmful if consumed frequently over time. Common additives include preservatives like sodium benzoate, which can exacerbate asthma, or artificial dyes linked to behavioral issues in children. Knowing which additives to avoid and understanding their effects on health is a critical skill in safeguarding your wellbeing.

For those looking to delve deeper into food additives, resources like the online database at the Center for Science in the Public Interest (CSPI) offer comprehensive guides that detail the uses, risks, and health impacts of various food additives. Barcode scanning apps like Fooducate allow consumers to scan products in real time and receive detailed health information and safer alternatives, facilitating informed decision-making while shopping.

Visual Element: Interactive Label Reading Exercise

To put this knowledge into practice, engage in an interactive label-reading exercise. Next time you shop, pick three products and scrutinize their labels using the above mentioned guidelines. Compare your findings and reflect on how the reality of the products compares to the marketing. This exercise reinforces your understanding and enhances your ability to make healthier and more ethical food choices consistently.

Armed with this knowledge, you step into the grocery store not just as a consumer but as an informed citizen, capable of making choices that align with your values and contribute to a more transparent and accountable food system. As you turn each page of this chapter and apply its lessons, remember that every product you put in your cart is a vote for the kind of world you want to live in—one where truth and health are cherished.

15.2 Tools for Tracking and Reporting Food Safety Violations

In our interconnected world, the power to influence food safety isn't just in the hands of regulatory bodies but also rests squarely in the palm of your hand—quite literally. Modern technology has equipped us with various online tools and mobile apps designed to empower you, the consumer, to monitor and report food safety violations actively. These digital platforms facilitate reporting and ensure your concerns are heard and acted upon effectively.

One of the most effective ways to leverage your consumer power is through mobile apps that allow you to report food safety issues directly from your smartphone. Apps like "Food Keeper," developed by the USDA, help you understand how to store food safely and provide a direct line to report issues. Another tool, "iwaspoisoned.com," enables individuals to report food borne illness cases, tracking and aggregating data to identify potential outbreaks in real time. These tools are crucial to providing clear, concise, and factual information. Include details such as the name and location of the establishment, the date and time of your visit, what you ordered, and the symptoms experienced, if any. By providing specific details, you help health authorities initiate investigations more efficiently and potentially prevent harm to others.

Engaging with local health departments is another critical avenue for addressing food safety concerns. Your local health department plays a pivotal role in inspecting food service establishments and responding to public health complaints. However, the reporting process can seem daunting if you're unfamiliar. Start by visiting your local health department's website, which typically guides how to report food safety issues. When making a report, be as specific as possible about your concerns and provide any evidence you have gathered, such as photos or receipts. Keeping a record of your communication with the department is also beneficial. Follow up if you are still waiting to receive a response within a reasonable timeframe, as this persistence often encourages quicker action.

Documentation is the backbone of effective food safety violation reporting. Taking clear photos can be incredibly impactful when encountering a potential safety violation. Photograph the product, focusing on any defects or contaminants, and the packaging details, including the batch number, expiry date, and manufacturer information. These details are crucial for traceability and investigation. Additionally, any purchase receipts or product packaging, as they serve as proof of purchase and further aid in the investigation process, should be retained. This thorough documentation strengthens your report and supports the investigative process, increasing the likelihood of a resolution.

Lastly, it's essential to consider the legal aspects of reporting food safety violations. In many jurisdictions, reporters of food safety violations are protected under whistleblower protection laws, ensuring you can raise concerns without fear of retaliation. However, it's essential to report in good faith and with honesty. False reports can undermine trust and waste valuable public health resources. If you need clarification on the legal implications of your report, consulting with legal experts or consumer rights organizations can provide clarity and confidence.

By embracing these tools and practices, you contribute to the safety and integrity of the food system and uphold your rights as a consumer. Through vigilant observation, meticulous documentation, and effective use of reporting channels, you can help ensure that food safety violations do not go unnoticed, fostering a safer dining environment for all.

15.3 Organizing Community Forums on Food Safety

In the quest to elevate food safety awareness and foster significant community engagement, organizing a community forum stands out as a powerful tool. These forums provide a platform for discussion, mobilize collective action, and shape future food safety initiatives. The key to a successful forum lies in meticulous planning and the effective engagement of stakeholders, ensuring that the event not only educates but also empowers participants.

Call To Action

Planning and Coordination

The first step in organizing a community forum on food safety involves detailed planning and coordination. Begin by selecting a venue accessible to all community members, including those with disabilities. This could be a community center, a school hall, or any public meeting space that can accommodate a large group equipped with necessary audio-visual facilities. Once the venue is secured, setting an agenda is crucial. This should include a clear outline of the topics to be discussed, the structure of the event, and the time allocated for each segment. Inviting speakers who are experts in food safety, public health, and related fields can provide valuable insights and draw a larger audience. These might include local health officials, nutritionists, or advocates who have been active in food safety campaigns. Additionally, ensure that the forum allows ample Q&A sessions, which are vital for addressing community concerns and fostering interactive dialogue.

Engaging Stakeholders

Engaging a diverse group of stakeholders is essential for a thriving community forum. This includes local businesses, health officials, community leaders, and the general public. Start by contacting local health departments and food safety organizations to invite them to participate as speakers or informational resources. Engaging local businesses, especially in the food industry, can be beneficial. They can provide practical insights into how food safety is managed at the ground level and what challenges they face. To encourage their participation, emphasize the benefits of improved community relationships and the opportunity to demonstrate their commitment to food safety. Community leaders and elected officials can help amplify the event's reach through their networks and influence, encouraging a more significant segment of the community to participate.

Facilitation Techniques

Effective facilitation ensures the forum remains focused, engaging, and inclusive. Establish ground rules for discussions to create a respectful and constructive environment. Use moderation techniques that encourage participation from all attendees, such as inviting questions or comments from those who might be less inclined to speak up. Simple strategies like round-robin speaking, where each person can speak without interruptions, can ensure a more inclusive dialogue. It's also beneficial to use visual aids and handouts summarizing critical information, as these can help attendees better understand complex topics and retain information long after the forum ends.

Post-Forum Action Steps

To maintain momentum post-forum, it's vital to establish clear follow-up actions. Forming action groups can channel the energy and ideas generated during the forum into concrete initiatives. These groups can work on specific issues, such as advocating for better food labeling practices or creating community gardens to improve local food security. Developing community initiatives, such as a regional food safety awareness week, can keep the conversation going and engage a broader audience. Additionally, setting regular meeting schedules for action groups ensures ongoing engagement and accountability, helping to sustain a long-term commitment to improving food safety in the community.

Organizing a community forum on food safety is not just about disseminating information; it's about building a community network to advocate for change and influence policies. Through careful planning, engaging diverse stakeholders, effective facilitation, and strategic follow-up, such forums can catalyze community action and empowerment, driving forward the collective mission of safer, healthier food for all.

15.4 Engaging Youth in Food Safety Advocacy

The seeds of change in food safety and advocacy are often sown in the fertile minds of the young, who bring energy, fresh perspectives, and a natural proficiency with new technologies that can amplify their voices. Engaging youth in food safety advocacy not only empowers them but also ensures the longevity and dynamism of the movement. One of the most effective ways to foster this engagement is through educational programs integrated into school curriculums. Collaborating with educators to develop programs encompassing the basics of food safety, the importance of nutrition, and the impact of food on the environment can cultivate an early awareness and interest in these critical issues. Such programs could include interactive workshops where students learn to read and understand food labels, recognize misleading food claims, and understand the global impact of their food choices. By engaging these sessions through multimedia presentations, hands-on activities, and guest speakers from the local food safety boards or health departments, students can connect theoretical knowledge with real-world applications, making the learning process both enjoyable and impactful.

Moreover, youth-led initiatives are potent platforms for young advocates to take charge and implement change. Across the globe, numerous success stories can inspire similar efforts. For instance, consider a high school where students campaigned to reduce food waste in their cafeteria. They conducted a waste audit, presented their findings to school administrators, and proposed solutions such as improved meal planning and composting programs. This initiative reduced waste and educated their peers about sustainable practices. Starting such a group can seem daunting, but with the right approach, it's entirely feasible. Begin by gathering a group of interested students and appointing a faculty advisor to provide guidance. Set clear, achievable goals, such as raising awareness about food additives or campaigning for healthier school lunches. Regular meetings and defined roles within the group can help maintain momentum while engaging in local community events can raise public awareness and support.

Social media platforms are invaluable tools for young advocates looking to raise awareness and campaign for food safety. Platforms like Instagram, TikTok, and Facebook offer vast networks where messages can quickly spread to local, national, and international audiences. Young people can use these platforms to share informative content, such as videos explaining the importance of food safety, infographics about how to avoid foodborne illnesses, or updates on their advocacy efforts. The key to success on social media lies in creating content that is not only informative but also relatable and engaging. Using a conversational tone, appealing visuals, and hashtags can increase the visibility of posts. Collaborations with influencers passionate about health and sustainability can further extend the reach of these messages, making them an integral part of the discourse on food safety.

Building leadership skills is crucial for young advocates, as these skills will empower them to lead initiatives effectively and inspire others to join their cause. Opportunities for leadership development can be found in workshops and mentorship programs offered by local community centers, non-profit organizations, and schools. These programs should focus on critical skills such as public speaking, project management, and strategic thinking. Networking events can also provide young people the chance to connect with experienced advocates and leaders in the field of food safety, offering them mentors who can guide their development and advise them on carrying out effective campaigns.

Through educational programs, youth-led initiatives, social media engagement, and leadership development, young people can become formidable advocates for food safety. Their involvement enriches their lives with valuable skills and experiences and ensures that the fight for safer, healthier food systems continues with renewed vigor and passion. As these young advocates grow, so does their potential to effect meaningful change, heralding a future where food safety is not just an ideal but a reality.

15.5 Writing to Legislators: A Template for Change

Navigating the legislative process can often seem like deciphering a complex puzzle where each piece represents a different stakeholder or procedural step. Understanding this process is crucial to effectively influencing food safety laws. Legislation typically starts as an idea or proposal, often inspired by public needs or advocacy groups. This proposal is then drafted into a bill by a legislator. Once introduced, the bill goes through various committees where it is reviewed, debated, and amended. Public hearings may be held, allowing you to voice support or concerns. If it passes through the committees, the bill is discussed on the floor of the legislative body. A vote is taken, and if passed, it moves to the other legislative chamber for a similar process. Upon approval by both chambers, the bill is sent to the governor or president, who can sign it into law or veto it. Each step in this process is an opportunity for advocacy, making it essential to know when and how to make your voice heard effectively.

Crafting effective letters to legislators is a foundational skill in the toolkit of any advocate. When writing to your representatives, clarity and conciseness are your allies. Start your letter by introducing yourself and stating your purpose briefly. For instance, you might begin with, "My name is [Your Name], a resident of [Your Town/City], and I am writing to express my concerns about [specific issue]." Immediately specify what action you want the legislator to take, such as supporting or opposing a bill. Include the bill number if applicable. Follow this with a brief but compelling argument as to why this issue matters, ideally personalizing the impact by sharing how it affects you, your family, or your community. This personal connection can transform your letter from a generic request to a compelling story that grabs attention. Conclude with a respectful call to action, reiterating what you ask the legislator to do and thanking them for their time and consideration. Remember, your goal is to persuade, not to lecture.

Petitions can serve as powerful tools in legislative advocacy, amplifying your voice through collective action. A well-organized petition shows legislators that an issue has broad support and is not just the concern of a few. When creating a petition, start with a clear and concise purpose statement explaining what you are petitioning for and why. This should be followed by a compelling narrative or data that supports your cause. Promoting your petition is crucial; utilize social media platforms, community groups, and local events to gather signatures. Once you have collected many signatures, please arrange to deliver the petition, ideally in person, during a meeting with your legislator or their staff. This provides an opportunity to discuss further the issue and the changes you hope to see.

Several case studies demonstrate that successful advocacy is often built on the foundation of well-crafted, targeted communications. Consider the case where a small community group successfully lobbied for stricter food labeling laws. They formed a coalition of concerned citizens and local health professionals, who drafted letters and petitions collectively. Regular meetings with their legislators, accompanied by expert testimonies on the health impacts of inadequate food labels, eventually led to the passage of new labeling standards that benefited consumers statewide. Another example involves a national campaign that used social media to raise awareness about the overuse of antibiotics in livestock. The campaign coordinated letter-writing blitzes and petitions, eventually leading to legislative changes that introduced stricter regulations on antibiotic use in farming.

In each case, combining personal stories, empirical evidence, and widespread public support was instrumental in swaying legislative opinion. As you write to your legislators, remember that your efforts are part of a larger tapestry of advocacy that can, over time, weave significant changes in public policy and food safety standards. Each letter, petition, and meeting contributes to a collective effort that has the power to reshape our food system for the better.

15.6 Future Technologies and Their Role in Consumer Empowerment

Imagine walking into a supermarket, pointing your smartphone at a product, and instantly receiving detailed information about its origin, ingredients, and associated health risks. This scenario isn't a glimpse into a distant future but a real possibility with today's emerging technologies like augmented reality (A.R.). A.R. apps can revolutionize how consumers interact with products by overlaying digital information onto real-world environments, providing transparency and detailed product insights at purchase. For instance, an A.R. app can display a visual breakdown of a product's nutritional content, flag potential allergens, or provide ethical ratings for the company based on its environmental and social practices. This instant access to comprehensive product data can shift consumer behavior towards more informed and health-conscious decisions.

Participatory monitoring systems represent another frontier in consumer empowerment, leveraging the collective power of community observations to ensure food safety. These systems utilize mobile and web-based platforms to allow consumers to enter real-time data about food quality and safety issues. By tapping into the widespread use of smartphones, such systems gather vast amounts of data from diverse geographic locations, enhancing the ability to detect and respond to food safety threats quickly. For example, a participatory monitoring app could allow users to report a spoiled product batch, triggering alerts to other consumers and prompting swift supplier investigations and regulatory responses. This collaborative approach enhances the responsiveness of food safety monitoring and fosters a community-oriented approach to public health.

Integrating blockchain technology into the food supply chain promises unprecedented transparency and traceability. By securely recording each step of a food product's journey from farm to table, blockchain creates a tamper-proof chain of custody that consumers can access. This means you could trace the origins of a steak back to the farm where the cattle were raised, verify that organic produce is genuinely organic, or confirm that seafood was sustainably harvested. The implications for food safety and ethical consumption are profound, as this level of transparency holds producers and suppliers accountable while giving consumers the confidence that the food they purchase meets their standards.

However, integrating these technologies is not without challenges and ethical considerations. Privacy concerns are paramount, as the collection of consumer data—ranging from purchasing habits to personal health information—poses significant risks. Ensuring this data is stored and used ethically requires robust cybersecurity measures and transparent data policies protecting consumer rights. Additionally, the digital divide remains a significant barrier, as not all consumers can access the necessary technology to benefit from these innovations. Addressing these disparities is crucial to prevent a scenario where only a fraction of the population can access the benefits of technological advancements in food safety.

As we look to the future, the role of technology in empowering consumers and enhancing food safety is clear. Innovations like A.R., participatory monitoring, and blockchain can transform the landscape of consumer rights and responsibilities, making the complex world of food supply chains open and accessible to everyone. By staying informed about these technologies and advocating for their responsible implementation, you can contribute to a future where technology advances our capabilities, upholds our values, and protects our health.

Visual Element: Infographic on Augmented Reality in Food Shopping

An infographic illustrates how A.R. technology can be used in supermarkets to provide consumers with instant product information. It highlights nutritional content, allergen alerts, and ethical ratings, showing how A.R. enhances shopping experiences and promotes informed consumer decisions.

Ending Note

As we close this chapter on future technologies and their role in consumer empowerment, we've explored how innovations like augmented reality, participatory monitoring, and blockchain could revolutionize our interaction with the food we consume. These technologies offer exciting possibilities for enhancing transparency, ensuring food safety, and empowering consumers with real-time, actionable information. As we continue to navigate the evolving landscape of food safety, integrating these technologies presents opportunities and challenges that will undoubtedly shape the future of our food systems. Moving forward, the next chapter will delve deeper into global initiatives and policies that are pivotal in shaping food safety standards and practices, ensuring that technological advancements align with broader goals of public health and sustainability.

CONCLUSION

As we draw the curtains on this enlightening journey through the complexities of our food safety systems, it's crucial to reflect on the critical issues we've navigated together. We've uncovered the hidden dangers lurking in our food—from toxic ingredients to bioengineered additives—and laid bare the flaws in a regulatory system that often puts industry interests above public health. We've also seen the power of global movements and the impact of informed, committed individuals and communities advocating for a safer food environment.

Throughout this book, we've stressed the importance of vigilance and knowledge. Understanding the pervasive presence of harmful additives and recognizing the gaps in our food safety regulations are foundational steps toward initiating change. Technology has emerged as a vital tool in improving food traceability and safety, empowering consumers and regulators alike to keep a closer watch on the integrity of our food.

Conclusion

The role of global advocacy and community action cannot be overstated. From local grassroots movements to international coalitions, collective efforts are reshaping policies and practices, steering us toward a more ethical and transparent food system. However, the journey continues. The path to reforming our food system is fraught with challenges, dominated by powerful interests resistant to change. It requires persistence, resilience, and an unwavering commitment to the cause.

I urge you, the reader, to not only digest the information shared but to act upon it. Engage with your community, question where your food comes from, demand clear labeling, and support policies prioritizing public health over corporate profits. Every small action contributes to a broader movement aimed at safeguarding our health and that of future generations.

Imagine a world where food safety is not an afterthought but a fundamental right, where regulatory bodies are not just watchdogs but guardians of public health. This vision is attainable if we stand united to demand and enact change.

Let us be the change we wish to see in our food system. Start today with one informed choice, one question, and one small step towards collective action. We can achieve a future where our food is nourishing, safe, and just for all.

Thank you for joining me on this journey. Let's continue pushing forward, learning, advocating, and inspiring change.

REFERENCES

How the FDA's food division fails to regulate health ... - Politico https://politico.com/interactives/2022/fda-fails-regulate-food-health-safety-hazards

How the FDA's food division fails to regulate health and ... https://politico.com/interactives/2022/fda-fails-regulate-food-health-safety-hazards

Health risks of genetically modified foods https://pubmed.ncbi.nlm.nih.gov/18989835/

Food Additives Amendment of 1958 https://en.wikipedia.org/wiki/Food_Additives_Amendment_of_1958

Artificial food additives: hazardous to long-term health - PubMed https://pubmed.ncbi.nlm.nih.gov/38423749/#:

How Safe are Color Additives? https://www.fda.gov/consumers/consumer-updates/how-safe-are-color-additives

Potential impacts of synthetic food dyes on activity and ... https://www.ncbi.nlm.nih.gov/pmc/articles/PMC9052604/

Potentials of Natural Preservatives to Enhance Food Safety ... https://www.ncbi.nlm.nih.gov/pmc/articles/PMC9525789/

Trans fat is double trouble for heart health - Mayo Clinic https://www.mayoclinic.org/diseases-conditions/high-blood-cholesterol/in-depth/trans-fat/art-20046114.

FDA Rules Corn Syrup Can't Change Its Name To Corn Sugar https://www.npr.org/sections/thesalt/2012/05/30/154009682/fda-rules-corn-syrup-cant-change-its-name-to-corn-sugar

A review of the alleged health hazards of monosodium ... https://www.ncbi.nlm.nih.gov/pmc/articles/PMC6952072/

Genetically modified foods: safety, risks and public concerns https://www.ncbi.nlm.nih.gov/pmc/articles/PMC3791249/

'Constant turmoil' at FDA's food regulatory agency, report ... https://www.politico.com/news/2022/12/06/fda-formula-report-foods-program-00072584.

Aspartame controversy - Wikipedia https://en.wikipedia.org/wiki/Aspartame_controversy#:

Experts Say Lobbying Skewed the U.S. Dietary Guidelines https://time.com/4130043/lobbying-politics-dietary-guidelines/

"Part of the Solution": Food Corporation Strategies for ... https://www.ncbi.nlm.nih.gov/pmc/articles/PMC9309978/

Food Regulations in Europe vs. the U.S. - Tilley Distribution https://www.tilleydistribution.com/food-regulations-in-europe-vs-the-us/

References

Ministry of Health, Labour and Welfare: Food https://www.mhlw.go.jp/english/topics/foodsafety/

Restrictions on Genetically Modified Organisms: Canada https://www.loc.gov/law/help/restrictions-on-gmos/canada.php

FDA Response to the Fukushima Daiichi Nuclear Power ... https://www.fda.gov/news-events/public-health-focus/fda-response-fukushima-daiichi-nuclear-power-facility-incident#:

New review links ultra-processed foods to 32 health problems https://www.washingtonpost.com/wellness/2024/02/29/ultraprocessed-foods-health-risk/#:

Common food additives and chemicals harmful to children https://www.health.harvard.edu/blog/common-food-additives-and-chemicals-harmful-to-children-2018072414326

The Epidemiology of Food Allergy in the Global Context https://www.ncbi.nlm.nih.gov/pmc/articles/PMC6163515/

Nutrition education has a significant impact on perceived ... https://www.ncbi.nlm.nih.gov/pmc/articles/PMC10755509/

Filing Whistleblower Complaints under the FDA Food Safety Modernization Act https://www.osha.gov/sites/default/files/publications/OSHA3714.pdf

50 Years Ago, Sugar Industry Quietly Paid Scientists To Point Blame At Fat https://www.npr.org/sections/thetwo-way/2016/09/13/493739074/50-years-ago-sugar-industry-quietly-paid-scientists-to-point-blame-at-fat

Genetically modified foods: safety, risks and public concerns—a review tips://www.ncbi.nlm.nih.gov/pmc/articles/PMC3791249/

Study: Chemicals found widespread in fast-food, groceries https://www.cbs8.com/article/news/local/harmful-chemicals-found-in-majority-of-food-and-fast-food/509-e9ecb96f-787c-4b44-912c-38ec61ec6fe3

How Journalists Are Investigating Food Stories Worldwide https://gijn.org/stories/from-source-to-table-how-journalists-are-investigating-food-stories-worldwide/

The media and food-risk perceptions - PMC https://www.ncbi.nlm.nih.gov/pmc/articles/PMC3128959/

The Dubious Influence of Social Media Trends on Food Safety https://www.food-safety.com/articles/9434-hashtags-and-hazards-the-dubious-influence-of-social-media-trends-on-food-safety

Misinformation and Disinformation in Food Science and Nutrition https://jn.nutrition.org/article/S0022-3166(22)13102-0/fulltext

10 of the Worst Food Safety Scandals in Recent History - Xtalks https://xtalks.com/10-of-the-worst-food-safety-scandals-in-recent-history-3435/

FOIA Success Stories https://www.justice.gov/sites/default/files/oip/legacy/2014/07/23/foia-success-stories.pdf

References

Organizations Committed to Food Advocacy and Changing ... https://foodmedcenter.org/8386-2/

Court smacks USDA for lack of transparency in GMO labeling https://www.thenewlede.org/2022/09/court-smacks-usda-for-lack-of-transparency-in-gmo-labeling/

Health risks of genetically modified foods https://pubmed.ncbi.nlm.nih.gov/18989835/

How Vegetable Oils Replaced Animal Fats in the American Diet https://www.theatlantic.com/health/archive/2012/04/how-vegetable-oils-replaced-animal-fats-in-the-American-diet/256155/

Big Food's Ambivalence: Seeking Profit and Responsibility ... https://www.ncbi.nlm.nih.gov/pmc/articles/PMC5296687/

International Regulations on Genetically Modified Organisms https://www.food-safety.com/articles/4826-international-regulations-on-genetically-modified-organisms-us-europe-china-and-japan

Health risks of genetically modified foods https://pubmed.ncbi.nlm.nih.gov/18989835/

Food Safety https://www.cspinet.org/advocacy/food-safety

EFSA Launches #EUChooseSafeFood Social Media Campaign in 16 Countries https://www.food-safety.com/articles/8589-efsa-launches-euchoosesafefood-social-media-campaign-in-16-countries

Widely consumed vegetable oil leads to an unhealthy gut https://www.universityofcalifornia.edu/news/widely-consumed-vegetable-oil-leads-unhealthy-gut

Development and Application of A.I. for Food Processing and Safety Regulations https://www.food-safety.com/articles/9387-development-and-application-of-ai-for-food-processing-and-safety-regulations

How Walmart brought unprecedented transparency to the ... https://www.hyperledger.org/case-studies/walmart-case-study

Recent advances in non-thermal processing technologies ... https://www.sciencedirect.com/science/article/pii/S277250222200213X

Focus on food safety and transparency https://www.foodsafetynews.com/2024/01/focus-on-food-safety-and-transparency/

Unexpected effects of urban food activism on community and ... https://www.tandfonline.com/doi/full/10.1080/13549839.2023.2298675

Mindfulness meditation: A research-proven way to reduce stress https://www.apa.org/topics/mindfulness/meditation

A Systematic Review of Organic Versus Conventional Food ... https://www.ncbi.nlm.nih.gov/pmc/articles/PMC7019963/

5 Ways to Avoid GMO or Bioengineered Food Ingredients https://rainbowacresca.com/blog/f/5-ways-to-avoid-gmo-or-bioengineered-food-ingredients

Understanding the Codex Alimentarius - Revised and Updated https://www.fao.org/4/y7867e/y7867e02.htm

References

Slow Food Movement Growing Fast - Our World https://ourworld.unu.edu/en/slow-food-movement-growing-fast.

ORGANIC FARMING :: Success Stories https://agritech.tnau.ac.in/ta/org_farm/orgfarm_success%20stories.html

Global Food Safety Initiative: Case study booklet https://nutritionconnect.org/resource-center/global-food-safety-initiative-case-study-booklet

How to Understand and Use the Nutrition Facts Label https://www.fda.gov/food/nutrition-facts-label/how-understand-and-use-nutrition-facts-label

Report a Problem with Food | FoodSafety.gov https://www.foodsafety.gov/food-poisoning/report-problem-with-food#:

Food Safety in Schools (formerly Serving It Safe) https://theicn.org/icn-resources-a-z/food-safety-in-schools/

Food Traceability Technologies: 7 Things to Know https://www.packagingdigest.com/food-safety/food-traceability-technologies-7-

Made in United States
Orlando, FL
11 November 2024